# DESIGN AND ANALYSIS OF EXPERIMENTS

# DESIGN AND ANALYSIS OF EXPERIMENTS

**Arjun K Gupta**
Bowling Green State University, USA

**D G Kabe**

NEW JERSEY • LONDON • SINGAPORE • BEIJING • SHANGHAI • HONG KONG • TAIPEI • CHENNAI

*Published by*

World Scientific Publishing Co. Pte. Ltd.

5 Toh Tuck Link, Singapore 596224

*USA office:* 27 Warren Street, Suite 401-402, Hackensack, NJ 07601

*UK office:* 57 Shelton Street, Covent Garden, London WC2H 9HE

**British Library Cataloguing-in-Publication Data**
A catalogue record for this book is available from the British Library.

**DESIGN AND ANALYSIS OF EXPERIMENTS**



ISBN 978-981-4522-53-3

Printed in Singapore

# Dedication

For Samik
The Indo-Irish Significant Boy
AKG

# Preface

Design of experiments holds a central place in statistics. The aim of this book is to present in a readily accessible form certain theoretical results of this vast field. This is intended as a text book for a one-semester or two-quarter course for undergraduate seniors or first-year graduate students or as a resource of supplementary reading and references. Basic knowledge of algebra, calculus and statistical theory is required to master the techniques presented in this book. We believe that students and most teachers of design of experiments will find the book both interesting and useful.

The book contains 9 chapters. Chapter 1 provides the basic statistical tools that are used in the remaining chapters. It can be read first or can be used as reference for readers who can begin with Chapter 2. The main body of the material is presented in Chapters 2–7. Chapter 8 is different in the sense that it is pure mathematics and provides the elements of modern algebra which are used in Chapter 9. The last chapter discusses the construction of designs. The topics in a book as this always come from many sources. We are grateful to many teachers, students and authors who directly or indirectly have contributed to this book. Our special thanks go to our graduate students, Hu and Ngoc for their help in proofreading the manuscript and for suggesting many improvements.

The authors would like to acknowledge the support of the Bowling Green State University and the St. Mary's University. The sad demise of Professor D. G. Kabe during the preparation of this volume has delayed the completion of the book. We are thankful to Ms. Lai Fun Kwong and the staff of World Scientific for their help and cooperation in bringing this book to publication. Finally, we would like to thank Lei Kou and Nathan Baxter whose skills at word processing made the preparation of this book far less painful than it might have been. However, any errors or omissions remain entirely the responsibility of the authors.

Arjun K. Gupta
D. G. Kabe

# Contents

Chapter 1

# Introduction and Preliminary Results

## 1.1 Introduction: Elementary Statistical Notions

The first concept in statistics is that of a population. By a population we mean the set of all possible values taken by a random variable. Such a population has a certain distribution which may be discrete or continuous according as the population is discrete or continuous. Any distribution is specified by a mathematical form containing some unknown quantities called parameters, and the distribution of a given population can be specified completely only when the values of these parameters are all known. These parameters will in general be functions of such measureable characteristics of the population like the mean, median, mode, variance etc. and the exact value of these quantities will be known only by a complete enumeration of the population which is a practical impossibility at least for infinite populations. So we restrict our study to samples which are chosen so as to represent the parental population adequately. The estimation of the unknown parameters from sample data is one of the most important functions of statistical theory. A second important function is the derivation of what may be called sampling distributions, that is, distributions of functions of sample values. These distributions will, of course, also be describable in terms of the parameters of the original distribution.

## 1.2 Sampling Distributions

### Normal and Related Distributions

The most important distribution in statistical theory is the normal distribution, $N(\mu, \sigma)$, which has the density function

$$f(x) = \frac{1}{\sqrt{2\pi}\,\sigma} e^{\frac{-(x-\mu)^2}{2\sigma^2}} ; -\infty < x < \infty.$$

In this case the parameters are $\mu$, the mean, and $\sigma > 0$, the standard deviation.

If $(x_1, x_2, \dots, x_n)$ is a random sample of size n from this population then the sample mean $\bar{x}$ follows the normal distribution $N(\mu, \frac{\sigma}{\sqrt{n}})$. The quantity $\frac{(n-1)s^2}{\sigma^2}$, where $s^2 = \frac{1}{n-1}\sum_{i=1}^{n}(x_i - \bar{x})^2$, has a chi-square distribution with $(n-1)$ degrees of freedom (d. f.). Also $\bar{x}$ and $s^2$ are independent. The quantity

$$t = \frac{(\bar{x} - \mu)}{s/\sqrt{n}} = \sqrt{n}(\bar{x} - \mu)/s$$

follows a Student's t-distribution with density

$$f(t) = \frac{\Gamma\left(\frac{n}{2}\right)}{\sqrt{(n-1)}\Gamma\left(\frac{1}{2}\right)\Gamma\left(\frac{n-1}{2}\right)} \frac{1}{\left(1+\frac{t^2}{n-1}\right)^{\frac{n}{2}}}, \quad -\infty < t < \infty.$$

Tests of significance regarding the difference of means of two normal populations are based on the Student's t-distribution.

If $x_1, x_2, \dots, x_k$ are k independent "standard normal variables", then the quantity

$$\chi_k^2 = x_1^2 + \dots + x_k^2$$

has a chi-square distribution with density function

$$f(\chi_k^2) = \frac{1}{2^{\frac{k}{2}}\Gamma\left(\frac{k}{2}\right)} e^{-\frac{\chi_k^2}{2}} (\chi_k^2)^{\frac{k}{2}-1}, \quad 0 \le \chi_k^2 < \infty.$$

If we consider a set of, say n, frequencies, then the quantity

$$\sum \frac{(0-e)^2}{e} = \left( \sum_{j=1}^{n} \frac{0_j^2}{e_j} - n \right)$$

where $0$ = observed frequency, $e$ = expected frequency, will have an approximate chi-square distribution with $(n-1)$ d.f.. The goodness of fit of a theoretical distribution can be tested using this quantity.

If $s_1^2$ and $s_1^2$ are two independent unbiased estimates of the variance of a normal distribution based $n_1$ and $n_2$ d.f.., (i.e., based samples of sizes $n_1 + 1$ and $n_2 + 1$) then the ratio $F = \frac{s_1^2}{s_2^2}$ has the F-distribution with $n_1$ and $n_2$ d.f.'s. The density function of F is

$$f(F) = \left(\frac{n_1}{n_2}\right)^{\frac{n_1}{2}} \frac{\Gamma\left(\frac{n_1+n_2}{2}\right)}{\Gamma\left(\frac{n_1}{2}\right)\Gamma\left(\frac{n_2}{2}\right)} \frac{F^{\frac{n_1}{2}-1}}{\left(1+\frac{n_1}{n_2}F\right)^{\frac{n_1+n_2}{2}}}, 0 \le F < \infty.$$

The F-distribution is used to compare the means of several normal distributions with the same variance or independent estimates of variances of a normal distribution. Now,

$$F = \frac{s_1^2}{s_2^2}$$

$$= \frac{\frac{1}{n_1}\sum_{i=1}^{n_1+1}(x_{1i}-\overline{x_1})^2}{\frac{1}{n_2}\sum_{i=1}^{n_2+1}(x_{2i}-\overline{x_2})^2}$$

$$= \left(\frac{n_2}{n_1}\right) \frac{\sum_{i=1}^{n_1+1}\left(\frac{x_{1i}-\overline{x_1}}{\sigma}\right)^2}{\sum_{i=1}^{n_2+1}\left(\frac{x_{2i}-\overline{x_2}}{\sigma}\right)^2},$$

$$= \frac{\chi_{n_1}^2/n_1}{\chi_{n_2}^2/n_2} \quad \text{where } \sigma^2 \text{ is the population variance.}$$

The F-ratio with $n_1$ and $n_2$ d.f. is usually denoted by $F(n_1, n_2)$.

Note that,

$$t^2 = \frac{n(\bar{x}-\mu)^2}{\frac{1}{n-1}\Sigma(x_i-\bar{x})^2} = \frac{\left(\frac{\bar{x}-\mu}{\frac{\sigma}{\sqrt{n}}}\right)^2}{\frac{1}{n-1}\Sigma\left(\frac{x_i-\bar{x}}{\sigma}\right)^2} = \frac{\frac{\chi_1^2}{1}}{\frac{\chi_{n-1}^2}{n-1}} = F(1, n-1).$$

## 1.3 Estimation Revisited

The main purpose of statistical techniques is to estimate properties of distributions and to test hypothesis about these properties. By the estimation of a property is meant the calculation from sample data of a quantity that will be taken as the value of the property for the population. There are, in general, many possible estimates of a property. In each of these cases, the function which is designed to estimate the property is called estimator and the value to which it gives rise in a particular case is called estimate. Estimation which deals with the estimation of certain unknown parameters in a known distribution is known as parameter estimation. We are interested in the estimators and not their estimates. The properties of a good estimator are the following:

(i) Unbiasedness
(ii) Consistency
(iii) Efficiency
(iv) Sufficiency.

We shall consider these properties one by one.

**(i) Unbiasedness**

Let us consider a one parameter family, say $f(x, \theta)$, where $\theta$ is the unknown parameter. Then an estimator $\hat{\theta}$ is said to be unbiased for $\theta$ if $E(\hat{\theta}) = \theta$. If $E(\hat{\theta}) > \theta$, then $\hat{\theta}$ is said to be positively biased and if $E(\hat{\theta}) < \theta$, then $\hat{\theta}$ is said to be negatively biased. Since

$$E\left(\frac{1}{n}\Sigma x_i\right) = \frac{1}{n}\Sigma E(x_i) = \frac{1}{n}\Sigma \mu = \mu,$$

the sample mean is an unbiased estimator of the population mean. Similarly, it can be seen that

$$s^2 = \frac{1}{n-1}\sum(x_i - \bar{x})^2, E(s^2) = \sigma^2,$$

is an unbiased estimator of the population variance, whereas the sample variance $\frac{1}{n}\sum(x_i - \bar{x})^2$ is not an unbiased estimator.

**(ii) Consistency**

An estimator $\hat{\theta}_n$, computed from a sample of size n, will be said to be consistent if , for any positive $\varepsilon$ and $\eta$, however small, these is an integer N such that

$$n \geq N \Rightarrow P_r\{|\hat{\theta}_n - \theta| < \varepsilon\} > 1 - \eta.$$

Thus $\hat{\theta}_n$ is a consistent estimator of $\theta$ if it converges to $\theta$ stochastically. It is easy to prove (using Chebyshev's inequality) that the sample mean is a consistent estimator of the population mean.

**(iii) Efficient**

If, for the two estimators $\hat{\theta}_1$ and $\hat{\theta}_2$, we have $V(\hat{\theta}_1) < V(\hat{\theta}_2)$ for all n, then $\hat{\theta}_1$ is said to be more efficient than $\hat{\theta}_2$. In the case of the sample mean and sample median, both are unbiased and consistent, but

$$V(\text{mean}) = \frac{\sigma^2}{n}, \quad V(\text{median}) = \frac{\pi\sigma^2}{2n},$$

where n is the sample size and $\sigma^2$ is the population variance. (Here the parent population is Normal.). Since $\frac{\pi}{2} = 1.57 > 1$, the sample mean is more efficient than the sample median.

An estimator is said to be efficient if it has the smallest possible variance.

**(iv) Sufficiency**

Let $(x_1, x_2, \ldots, x_n)$ be a random sample of size n from a population $f(x, \theta)$. The joint p.d.f. of the sample is the likelihood function

$$L(\theta) = \prod_{i=1}^{n} f(x_i; \theta).$$

For a given sample this is a function of θ and regarded as a function of θ, L is called the likelihood function of θ. Let $\hat{\theta}$ be an estimator of θ. Then $\hat{\theta}$ is said to be a sufficient estimator of θ if we can express L in the form

$$L(\theta) = L_1(\hat{\theta}, \theta) \cdot L_2(x_1, x_2, \ldots, x_n)$$

where $L_1$ is a function of $\hat{\theta}$ and θ only and $L_2$ is independent of θ.

Consider a sample of size n from $N(\theta, \sigma)$. The likelihood function is

$$\begin{aligned} L(\theta) &= \left(\frac{1}{2\pi}\right)^{\frac{n}{2}} \frac{1}{\sigma^n} e^{-\frac{1}{2\sigma^2}\sum_{j=1}^{n}(x_j-\theta)^2} \\ &= \left(\frac{1}{\sqrt{2\pi}\sigma}\right)^{n} e^{-\frac{n}{2\sigma^2}(\bar{x}-\theta)^2} e^{-\frac{1}{2\sigma^2}\sum_{j=1}^{n}(x_j-\bar{x})^2} \\ &= L_1(\bar{x}, \theta) \cdot L_2(x_1, x_2, \ldots, x_n), \end{aligned}$$

and hence $\bar{x}$ is a sufficient estimator of θ.

**General Case**

Consider a population with density function $f(x; \theta_1, \theta_2, \ldots, \theta_k)$ where the θ's are k unknown parameters. Let $\hat{\theta}_1, \ldots, \hat{\theta}_k$ be a set of estimators. Then this set of estimators is said to be

(i) Unbiased if $E(\hat{\theta}_i) = \theta_i$ for all i

(ii) Consistent if each $\theta_i$ is consistent

(iii) Efficient if each $\theta_i$ is efficient, and

(iv) Sufficient if the likelihood function is expressed in the form.

$$L(\theta_1, \ldots, \theta_k) = L_1(\hat{\theta}_1, \ldots, \hat{\theta}_k; \theta_1, \ldots, \theta_k) \cdot L_2(x_1, x_2, \ldots, x_n),$$

where $L_1$ is a function of $\hat{\theta}_1, \dots, \hat{\theta}_k$ and $\theta_1, \dots, \theta_k$ only and $L_2$ is independent of $\theta_1, \dots, \theta_k$.

## 1.4 Methods of Estimation

### (i) Method of Maximum Likelihood

Consider the likelihood function $L = \prod_{i=1}^{n} f(x_i; \theta)$ of a random sample of size n from a population $f(x; \theta)$. Regarding L as a function of $\theta$, the principle of maximum likelihood is to select that value of $\theta$ which maximizes L as an estimate of $\theta$. In other words, the maximum likelihood estimator $\hat{\theta}$ is the solution, if any, of the equation $\frac{\partial L}{\partial \theta} = 0$ which satisfies the condition $\frac{\partial^2 L}{\partial \theta^2} < 0$. In practice it is convenient to maximize $\log L$ instead of L and so we take $\hat{\theta}$ as the solution of $\frac{\partial \log L}{\partial \theta} = 0, \frac{\partial^2 \log L}{\partial \theta^2} < 0$.

If the frequency function of the parent population is $f(x_i; \theta_1, \theta_2, \dots, \theta_k)$, then we maximize L with respect to $(\theta_1, \theta_2, \dots, \theta_k)$, where

$$L = \prod_{i=1}^{n} f(x_i; \theta_1, \theta_2, \dots, \theta_k).$$

**Example 1.1**

For the normal population $N(\mu, \sigma)$, the likelihood function is

$$L = \left(\frac{1}{\sqrt{2\pi}\sigma}\right)^n e^{-\frac{1}{2\sigma^2}\sum_{i=1}^{n}(x_i-\mu)^2}$$

$$\log L = -n\log\left(\sqrt{2\pi}\right) - n\log\sigma - \frac{1}{2\sigma^2}\sum_{i=1}^{n}(x_i - \mu)^2.$$

The maximum likelihood estimating equations are

$$\frac{\partial \log L}{\partial \mu} = 0, \quad \frac{\partial \log L}{\partial \sigma} = 0.$$

$\frac{\partial \log L}{\partial \mu} = 0$ gives $\frac{1}{\sigma^2}\sum(x_i - \mu) = 0$, the solution of which is $\mu = \bar{x}$.

Again, $\frac{\partial \log L}{\partial \sigma} = 0$ gives $-\frac{n}{\sigma} + \frac{1}{\sigma^3}\sum(x_i - \mu)^2 = 0$. The solution is $\sigma^2 = s^2$, where $s^2 = \frac{1}{n}\sum_{i=1}^{n}(x_i - \bar{x})^2$.

Thus the sample mean $\bar{x}$ and the sample variance $s^2$ are the maximum likelihood estimators of $\mu$ and $\sigma^2$.

Now, $E(\bar{x}) = \mu$ and $E(s^2) = \left(\frac{n-1}{n}\right)\sigma^2$. Thus $\bar{x}$ is unbiased for $\mu$ whereas $s^2$ is a biased estimator for $\sigma^2$.

**(ii) Method of Least Squares**

The method of least squares consists in taking as estimates of the parameters those values of the parameters that minimize the sum of square of the deviations of the actual values from their expected values, i.e., by minimizing $\sum[x_i - E(x_i)]^2$.

## 1.5 Interval Estimation

Instead of attempting to estimate $\theta$ by a function which, for a specific sample, gives a unique number (point estimation), we shall now consider the specification of a range for $\theta$. This is called interval estimation. Based on a sample we will give an interval such that the population parameter in expected to lie in that interval with a certain probability. This interval is called the confidence interval and the limits of this interval are called confidence limits.

**Example 1.2**

Consider the problem of setting up a confidence interval for the mean of a normal population. If $(x_1, \dots, x_n)$ is a random sample of size n

from $N(\mu, \sigma)$, then the sample mean $\bar{x}$ is an unbiased estimator of $\mu$ and has distribution $N\left(\mu, \frac{\sigma}{\sqrt{n}}\right)$.
But then

$$\Pr.\left[-1.96 \leq \frac{\bar{x} - \mu}{\frac{\sigma}{\sqrt{n}}} \leq 1.96\right] = 0.95,$$

equivalently

$$\Pr.\left[\bar{x} - 1.96\frac{\sigma}{\sqrt{n}} \leq \mu \leq \bar{x} + 1.96\frac{\sigma}{\sqrt{n}}\right] = 0.95.$$

Thus if $\sigma$ is known, $(\bar{x} - 1.96\frac{\sigma}{\sqrt{n}}, \bar{x} + 1.96\frac{\sigma}{\sqrt{n}})$ is a 95% confidence interval for $\mu$. If $\sigma$ is unknown, then for large sample size an approximate 95% confidence interval for $\mu$ can be obtained by replacing $\sigma$ by its estimate s, the sample standard deviation. When $\sigma$ is unknown and the sample size is small, an exact 95% confidence interval for $\mu$ can be constructed using the t-distribution. In this case the quantity $t = \frac{\bar{x}-\mu}{s/\sqrt{n-1}}$ has Student's t-distribution with $(n - 1)$ d.f. and a 95% confidence interval for $\mu$ is

$$\left(\bar{x} - t_{0.025}\frac{s}{\sqrt{n-1}},\ \bar{x} + t_{0.025}\frac{s}{\sqrt{n-1}}\right)$$

where $t_{0.025}$ is the 2.5% upper percentile of t with $(n - 1)$ d.f..

## 1.6 Linear Functions of Normally Distributed Variables

If $x_1, \ldots, x_n$ are independent normal variates with means $\mu_1, \ldots, \mu_n$ and variances ${\sigma_1}^2, \ldots, {\sigma_n}^2$, say, then any linear function of the form

$$\lambda_1 x_1 + \cdots + \lambda_n x_n$$

is normally distributed with mean $\lambda_1\mu_1 + \cdots + \lambda_n\mu_n$ and variance $\lambda_1^2{\sigma_1}^2 + \cdots + \lambda_n^2{\sigma_n}^2$,

If $\bar{x}$ is the mean of a random sample $(x_1, \dots, x_n)$ from a normal distribution of mean $\mu$ and variance $\sigma^2$, then $\bar{x} = \frac{1}{n}\sum x_i$ and $\mu_1 = \dots = \mu_n = \mu$, $\lambda_1 = \dots = \lambda_n = \frac{1}{n}$, so that $\bar{x}$ is normally distributed with mean $\mu$ and variance $\frac{\sigma^2}{n}$. Substituting $s^2 = \frac{1}{n-1}\sum_{i=1}^{n}(x_i - \bar{x})^2$ for $\sigma^2$, we have that $\frac{\bar{x}-\mu}{s/\sqrt{n-1}}$ is distributed as a Student's t with $(n-1)$ d.f..

Again, consider the difference between the mean $\bar{x}_1$ of a sample of size $n_1$ from a population with variance ${\sigma_1}^2$ and the mean $\bar{x}_2$ of a sample of size $n_2$ from a population with variance ${\sigma_2}^2$. Then

$$\begin{aligned} V(\bar{x}_1 - \bar{x}_2) &= V(\bar{x}_1) + V(\bar{x}_2) - 2\mathrm{Cov}(\bar{x}_1, \bar{x}_2) \\ &= V(\bar{x}_1) + V(\bar{x}_2), \text{ if } \bar{x}_1 \text{ and } \bar{x}_2 \text{ are independent.} \end{aligned}$$

If the variances are the same, say $\sigma^2$, then

$$V(\bar{x}_1 - \bar{x}_2) = \sigma^2\left(\frac{1}{n_1} + \frac{1}{n_2}\right).$$

From this we can derive the t-test for the difference $(\mu_1 - \mu_2)$. When $\sigma^2$ is unknown, the statistic

$$t = \frac{\bar{x}_1 - \bar{x}_2}{\sqrt{\sum_{i=1}^{n_1}(x_{1i} - \bar{x}_1)^2 + \sum_{j=1}^{n_2}(x_{2j} - \bar{x}_2)^2}}\sqrt{\frac{(n_1 + n_2 - 2)n_1 n_2}{n_1 + n_2}}$$

is distributed as Student's t with $(n_1 + n_2 - 2)$ d.f.. When $\sigma^2$ is known, $\frac{\bar{x}_1 - \bar{x}_2}{\sigma\sqrt{\frac{1}{n_1}+\frac{1}{n_2}}}$ is distributed as $N(0,1)$ and can be used to test the difference $(\mu_1 - \mu_2)$.

## 1.7 Cochran's Theorem

Let $y_1, \dots, y_n$ be n independent random variables all distributed according to $N(0,1)$, and let $Q_1, \dots, Q_k$ be quadratic forms in $y_1, \dots, y_n$ such that

$$\sum_{i=1}^{n} {y_i}^2 = Q_1 + \dots + Q_k.$$

Let $r_i = \text{rank}\, Q_i, i = 1,2, \ldots, k$. Then the necessary and sufficient condition that $Q_1, \ldots, Q_k$ follow independent $\chi^2$-distributions with $r_1, \ldots, r_k$ d. f. is that $n = \sum_{i=1}^{k} r_i$.

**Proof.**

If $Q_1, \ldots, Q_k$ are independent $\chi^2$'s with $r_1, \ldots, r_k$ d.f., respectively, then by the additive property of $\chi^2$'s it follows that $\sum_{i=1}^{k} Q_i$ is a $\chi^2$ with $\sum_{i=1}^{k} r_i$ d.f.. But $\sum_{i=1}^{k} Q_i = \sum_{i=1}^{n} {y_i}^2$ which is a $\chi^2$ with n d.f.. Hence $\sum_{i=1}^{k} r_i = n$, showing that the condition is necessary.

Conversely, let $\sum_{i=1}^{k} r_i = n$. Then there exists an orthogonal transformation $y = Cx$ which changes each $Q_i$ into a sum of squares according to the relations:

$$Q_1 = \sum_{i=1}^{r_1} {x_i}^2, Q_2 = \sum_{r_1+1}^{r_1+r_2} {x_i}^2, \ldots, Q_k = \sum_{n-r_k+1}^{n} {x_i}^2 .$$

Now, the joint distribution of $y_1, \ldots, y_n$ is

$$\text{constant} \times e^{-\frac{1}{2}y^T y} dy_1 \ldots dy_n,$$

and therefore the joint distribution of $x_1, \ldots, x_n$ is

$$\text{constant} \times e^{-\frac{1}{2}x^T x} dx_1 \ldots dx_n$$

showing that $x_1, \ldots, x_n$ are independent normal variates all following $N(0,1)$. Therefore, $Q_1, \ldots, Q_k$ are independent $\chi^2$'s with $r_1, \ldots, r_k$ d.f., respectively. ■

# Chapter 2

# Theory of Linear Estimation

## 2.1 Basic Assumptions and Definition of Least Squares

Let y be a row vector of n independent stochastic variables $y_1, \dots, y_n$ with

$$E(y) = \theta A^T \text{ (Observational Equations)}$$

where $\theta = (\theta_1, \dots, \theta_k)$ is a vector of k unknown parameters and $A = (a_{ij})$ is an $n \times k$ matrix with known elements. Let us also assume that the $y_i$'s have a common, but not necessarily known, variance $\sigma^2$, so that the covariance matrix of y is given by

$$V(y) = \sigma^2 I$$

where I stands for the identity matrix of order n.

A linear parameter function $C\theta^T = c_1\theta_1 + \cdots + c_k\theta_k$ with known coefficient row vector C is said to be estimable if there exists a linear function

$$By^T = b_1 y_1 \dots + b_n y_n$$

such that $E(By^T) = C\theta^T$ and such a function, if it exists, is called an unbiased estimator of $C\theta^T$. If no such function exists, the parameter function is said to be non-estimable.

An unbiased estimator with the minimum possible variance is said to be the best linear unbiased estimator. The mathematical discussion of arriving at an estimate with the minimum possible variance out of a large class of unbiased estimators is known as the theory of linear estimation.

**Theorem 2.1**

The necessary and sufficient condition for the estimability of a linear parameter function $C\theta^T$ is that the vector C should belong to the range space of the matrix A, i.e. , the matrices A and $A_0 = \binom{A}{C}$ (got by augmenting A with the row C) have the same rank.

**Proof.**

Necessity:

If $By^T$ is an unbiased estimator of $C\theta^T$, then $E(By^T) = C\theta^T$. But

$$\begin{aligned} E(By^T) &= B \cdot E(y^T) \\ &= BA\theta^T. \end{aligned}$$

Therefore the condition is $BA\theta^T = C\theta^T$. This must be true for all $\theta$. Hence we must have $BA = C$ which shows that the vector C belongs to the row space of A. So the condition is necessary.

Sufficiency:

Let C belong to the row space of A. Then the equation $BA = C$ always admits a non-trivial solution for the unknown B. For such a solution B, we have

$$BA\theta^T = C\theta^T, \text{or } E(By^T) = C\theta^T.$$

Hence there exists a linear function $By^T$ such that $E(By^T) = C\theta^T$. Hence $C\theta^T$ is estimable. This establishes the sufficiency of the condition. ■

Thus when $C\theta^T$ is estimable the coefficient vector B of the unbiased estimator $By^T$ must satisfy the equation $BA = C$. But the equation may have an infinity of solutions in B giving rise to an infinity of unbiased estimators of $C\theta^T$. If the solution is unique then there is only one unbiased estimator and that is the best possible. In the case of more than one solution, the problem of selecting the best estimator can be solved by imposing the natural restriction that the variance of such an estimator should be the least. Now,

$$\begin{aligned} V(By^T) &= B \cdot Cov(y^T, y) \cdot B^T \\ &= BB^T\sigma^2, \ \text{since } Cov(y^T, y) = \sigma^2 \cdot I \end{aligned}$$

Also when $C\theta^T$ is estimable, the relation $BA = C$ should be invariably satisfied. Since $\sigma^2$ is a constant, the problem row reduces to that of minimizing $BB^T$ subject to the condition $BA = C$. To obtain this restricted minimum, let us adopt the method of Lagrangian multipliers.

Let $\lambda = (\lambda_1, \lambda_2, \dots, \lambda_k)$ be a Lagrangian multiplier. Consider the expression

$$L = BB^T - 2\lambda(BA - C)^T.$$

The minimizing equations are

$$\frac{\partial L}{\partial B} = 2B - 2\lambda A^T = 0$$

and

$$\frac{\partial L}{\partial \lambda} = (BA - C)^T = 0.$$

i.e., $B = \lambda A^T$ and $BA = C$. The second equation is the same as the given condition. We have to solve for $\lambda$ using the two equations.

Since $B = \lambda A^T$, we have

$$C = BA = \lambda A^T A$$

which gives $\lambda$, and then the best estimate is given by

$$By^T = \lambda A^T y^T.$$

The equation $C = \lambda A^T A$ gives, in general, more than one solution for $\lambda$. But we shall prove that the estimator $By^T = \lambda A^T y^T$ is unique whatever the solution for $\lambda$, so that it is enough to get a simple solution for $\lambda$.

(Note: Since $C\theta^T$ is assumed to be estimable, A and $A_0 = \binom{A}{C}$ have the same rank. But then $A^T A$ and $\binom{A^T A}{C}$ have the same rank, and consequently, $C = \lambda A^T A$ is consistent.)

Let $\lambda_1$ and $\lambda_2$ be two solutions so that

$$\lambda_1 A^T A = C \quad \text{and} \quad \lambda_2 A^T A = C$$

$$\text{or} \quad (\lambda_1 - \lambda_2) A^T A = 0.$$

Consider the new quantity $P = (\lambda_1 - \lambda_2) A^T y^T$. Then

$$\begin{aligned} E(P) &= (\lambda_1 - \lambda_2)A^T E(y^T) \\ &= (\lambda_1 - \lambda_2)A^T A\theta^T \\ &= 0, \text{ since } (\lambda_1 - \lambda_2)A^T A = 0. \end{aligned}$$

$$\begin{aligned} \text{Again, } V(P) &= (\lambda_1 - \lambda_2)A^T \cdot \text{cov}(y^T, y) \cdot A(\lambda_1 - \lambda_2)^T \\ &= (\lambda_1 - \lambda_2)A^T A(\lambda_1 - \lambda_2)^T \sigma^2 \\ &= 0, \text{ since } (\lambda_1 - \lambda_2)A^T A = 0. \end{aligned}$$

Thus P has zero mean and zero variance. Consequently $P = 0$. But this means

$$\lambda_1 A^T y^T = \lambda_2 A^T y^T,$$

showing the uniqueness of $\lambda A^T y^T$.

Thus $By^T = \lambda A^T y^T$ is the unique unbiased estimator having the minimum variance, where $\lambda$ is a solution of $C = \lambda A^T A$.

**Theorem 2.2 (Gauss-Markoff)**

The unique best unbiased estimator of any estimable parameter function $C\theta^T$ is obtained by substituting for $\theta$ any solution of the least square normal equation in $\theta$ obtained by minimizing

$$(y - \theta A^T)(y - \theta A^T)^T \text{ w.r.t. } \theta.$$

**Proof.**

The error sum of squares is given by

$$L = (y - \theta A^T)(y - \theta A^T)^T,$$

which has to be minimized w.r.t. $\theta$. Differentiating w.r.t. $\theta$ and equating the derivative to 0, we get

$$yA = \theta A^T A.$$

These are called the normal equations. They are always consistent as is shown below.

We have Rank $A^T$ = Rank $(A^T, yA)$ and hence Rank $(A^TA)$ = Rank $(A^TA, yA)$. But this implies the consistency of the system of normal equations. If $A^TA$ is non-singular, the system has a unique solution given by

$$\hat{\theta} = yA(A^TA)^{-1}.$$

In any case, let us denote a solution by $\hat{\theta}$.

Suppose now that $C\,\theta^T$ is estimable so that there exists a $\lambda$ for which $C = \lambda A^TA$. Then

$$\begin{aligned} C(\hat{\theta})^T &= \lambda A^TA(\hat{\theta})^T \\ &= \lambda(\hat{\theta}A^TA)^T \\ &= \lambda(yA)^T \\ &= \lambda A^Ty^T \end{aligned}$$

which proves the assertion of the theorem. ■

This property is known as the Principle of Substitution.

**Note:**

Only a single solution is sufficient for the purpose of substitution. In many practical situations the normal equations have unique solutions, in which case all parameter functions are uniquely estimable.

The variance of the best estimator is given by

$$\begin{aligned} V(\lambda A^Ty^T) &= \lambda A^T \cdot \mathrm{Cov}(y^T, y)A\lambda^T \\ &= \lambda A^TA\lambda^T\sigma^2, \text{ since } \mathrm{Cov}(y^T, y) = \sigma^2 I \\ &= C\lambda^T\sigma^2, \text{ since } C = \lambda A^TA. \end{aligned}$$

Here C and $\lambda$ are known, but $\sigma^2$ is unknown. $C\lambda^T$ is called the relative variance of $By^T$.

In the particular case when $(A^TA)^{-1}$ exists, the solution for $\theta$ is uniquely given by

$$\hat{\theta} = yA(A^TA)^{-1}.$$

Let us find the variance of this estimator for $\theta$.

$$\begin{aligned}
V(\hat{\theta}) &= V\left(yA(A^TA)^{-1}\right) \\
&= \left(A(A^TA)^{-1}\right)^T \mathrm{Cov}(y^T, y)A(A^TA)^{-1} \\
&= \left((A^TA)^{-1}\right)^T A^T\mathrm{Cov}(y^T, y)A(A^TA)^{-1} \\
&= \left((A^TA)^T\right)^{-1} A^TA(A^TA)^{-1}\sigma^2 \\
&= (A^TA)^{-1}(A^TA)(A^TA)^{-1}\sigma^2 \\
&= (A^TA)^{-1}\sigma^2.
\end{aligned}$$

Therefore $(A^TA)^{-1}$ gives the relative variance of $\hat{\theta}$. Again,

$$V(\hat{\theta}) = (A^TA)^{-1}\sigma^2$$

$$= \begin{pmatrix} V(\hat{\theta}_1) & \mathrm{Cov}(\hat{\theta}_1, \hat{\theta}_2) & \cdots & \mathrm{Cov}(\hat{\theta}_1, \hat{\theta}_k) \\ \mathrm{Cov}(\hat{\theta}_1, \hat{\theta}_2) & V(\hat{\theta}_2) & \cdots & \mathrm{Cov}(\hat{\theta}_2, \hat{\theta}_k) \\ \cdots & \cdots & \cdots & \cdots \\ \mathrm{Cov}(\hat{\theta}_1, \hat{\theta}_k) & \mathrm{Cov}(\hat{\theta}_2, \hat{\theta}_k) & \cdots & V(\hat{\theta}_k) \end{pmatrix}$$

Thus the (i,j)-th element of $(A^TA)^{-1}$ gives $\frac{\mathrm{Cov}(\hat{\theta}_i,\hat{\theta}_j)}{\sigma^2}$ and the i-th diagonal element gives $\frac{V(\hat{\theta}_i)}{\sigma^2}$.

Further if $C_1\theta^T$ and $C_2\theta^T$ are two estimable parameter functions with best estimates $\lambda_1A^Ty^T$ and $\lambda_2A^Ty^T$ respectively, then their covariance is

$$\begin{aligned}
&\mathrm{Cov}\left(\lambda_1 A^T y^T, \lambda_2 A^T y^T\right)\\
&= \lambda_1 A^T \mathrm{Cov}\left(y^T, y\right) A \lambda_2{}^T\\
&= \lambda_1 A^T A \lambda_2{}^T \sigma^2\\
&= C_1 \lambda_2^T \sigma^2 \quad \text{or} \quad \lambda_1 C_2^T \sigma^2
\end{aligned}$$

since $C_1 = \lambda_1 A^T A$ and $C_2 = \lambda_2 A^T A$.

**Example 2.1**

Let $y = (y_1, y_2, y_3, y_4)$, $\theta = (\theta_1, \theta_2, \theta_3)$ and

$$A = \begin{pmatrix} 1 & 2 & 1 \\ 2 & 0 & 1 \\ 4 & 4 & 3 \\ 1 & -2 & 0 \end{pmatrix}.$$

Let the parametric function to be estimated is $C\theta^T$ where $C = (5,6,4)$.

$C\theta^T$ is estimable as can be easily checked. Rank $A = 2$ and so there are 2 independent linear estimable parameter functions, one of which is already given by $C\theta^T$. Therefore there is one more independent linear estimable parameter function.

Now, $A^T A = \begin{pmatrix} 22 & 16 & 15 \\ 16 & 24 & 14 \\ 15 & 14 & 11 \end{pmatrix}$.

Hence the normal equations are

$$(y_1, y_2, y_3, y_4) \begin{pmatrix} 1 & 2 & 1 \\ 2 & 0 & 1 \\ 4 & 4 & 3 \\ 1 & -2 & 0 \end{pmatrix} = (\theta_1, \theta_2, \theta_3) \begin{pmatrix} 22 & 16 & 15 \\ 16 & 24 & 14 \\ 15 & 14 & 11 \end{pmatrix}$$

There are three equations which are not independent since Rank $A = 2$. Therefore we can impose any one condition and solve. Let us put $\theta_1 = 0$. Then the first two equations become

$$16\theta_2 + 15\theta_3 = T_1,$$

$$24\theta_2 + 14\theta_3 = T_2,$$

where $T_1 = y_1 + 2y_2 + 4y_3 + y_4$, and $T_2 = 2y_1 + 4y_3 - 2y_4$. Solving for $\theta_2$ and $\theta_3$ we get

$$\theta_2 = \frac{15T_2 - 14T_1}{136}, \theta_3 = \frac{24T_1 - 16T_2}{136}.$$

Thus

$$\hat{\theta} = (0, \frac{15T_2 - 14T_1}{136}, \frac{24T_1 - 16T_2}{136})$$

and the unique best unbiased estimator of $C\theta^T$ is

$$\begin{aligned} C(\hat{\theta})^T &= 5 \times 0 + 6\left(\frac{15T_2 - 14T_1}{136}\right) + 4\left(\frac{24T_1 - 16T_2}{136}\right) \\ &= \frac{6T_1 + 13T_2}{68} \\ &= \frac{32y_1 + 12y_2 + 76y_3 - 20y_4}{68}. \end{aligned}$$

If we substitute any other solution of the normal equations, the estimator will be found to be the same thus verifying the uniqueness property of the best unbiased estimator.

The Langrangian multiplier $\lambda$ is given by

$$C = \lambda A^T A.$$

$$\text{i.e.}, (5,6,4) = (\lambda_1, \lambda_2, \lambda_3)\begin{pmatrix} 22 & 16 & 15 \\ 16 & 24 & 14 \\ 15 & 14 & 11 \end{pmatrix}.$$

Only two of the equations are independent. One solution is given by

$$\lambda = \left(0, \frac{5}{34}, \frac{3}{17}\right).$$

Therefore the variance of the estimator is

$$C\lambda^T\sigma^2 = (5,6,4)\begin{pmatrix} 0 \\ \frac{5}{34} \\ \frac{3}{17} \end{pmatrix}\sigma^2 = \frac{27}{17}\sigma^2.$$

Thus the relative variance of the estimator is $\frac{27}{17}$.

## 2.2 Linear Functions with Zero Expectation

We have seen that a linear parametric function $C\theta^T$ is estimable if and only if C belongs to the row space of A. Therefore, if r is the rank of A, then there will be r independent linear estimable parametric functions.

A linear function $By^T$ is said to be an error function if $E(By^T) = 0$ for all $\theta$. Since $E(By^T) = BA\theta^T$, for $By^T$ to be an error function one must have

$$BA = 0,$$

a necessary and sufficient condition for which is that the vector B is orthogonal to the row space $V_1$ of the matrix A, that is, C should belong to the orthogonal complement $V_0$ of $V_1$ regarded as a subspace of the k-vector space. If A is of rank n, then $V_1$ will be of dimension r and consequently $V_0$ will be of dimension k-r. We call $V_1$ the estimation space and $V_0$ the error space. The appropriateness of these names becomes clear from the following:

**Theorem 2.3**

If $C\theta^T$ is an estimable parametric function, then there exists a unique vector $B_1$, of the estimation space such that

$$E(B_1y^T) = C\theta^T.$$

Further, this linear function $B_1y^T$ is the best estimator of $C\theta^T$.

**Proof.**

Since $C\theta^T$ is estimable there exists a linear function $By^T$ such that $E(By^T) = C\theta^T$. Let $B_1$ be the projection of B on $V_1$ and $B_0$ its projection on $V_0$. Then $B = B_1 + B_0$ which is a unique representation. Now,

$$E(By^T) = E(B_1y^T + B_0y^T)$$

$$= E(B_1 y^T), \text{ since } E(B_0 y^T) = 0.$$

Therefore

$$E(B_1 y^T) = C\theta^T. \blacksquare$$

**Corollary 2.3.1**

If we know any linear unbiased estimator of an estimable parametric function $C\theta^T$, then we can find the best unbiased estimator of $C\theta^T$ at once, remembering that if B is the coefficient vector of this unknown estimator then the coefficient vector $B_1$ of the best estimator is simply the projection of B on the estimation space.

**Corollary 2.3.2**

The correspondence between an estimable parametric function and its best estimator is one-to-one. If $B_1 y^T, \ldots, B_m y^T$ are the best estimators of $C_1\theta^T, \ldots, C_m\theta^T$, then

$$By^T = \alpha_1 B_1 y^T + \cdots + \alpha_m B_m y^T$$

$$= (\alpha_1 B_1 + \cdots + \alpha_m B_m) y^T$$

is the best estimator of

$$C\theta^T = \alpha_1 C_1 \theta^T + \cdots + \alpha_m C_m \theta^T$$

$$= (\alpha_1 C_1 + \cdots + \alpha_m C_m)\theta^T$$

where $\alpha_1, \ldots, \alpha_m$ are scalars . For, evidently,

$$E(By^T) = E(\alpha_1 B_1 y^T + \cdots + \alpha_m B_m y^T)$$

$$= \alpha_1 C_1 \theta^T + \cdots + \alpha_m C_m \theta^T$$

$$= C\theta^T.$$

Also, since $B_1, \ldots, B_m$ lie in the estimation space, the linear combination $B = \alpha_1 B_1 + \cdots + \alpha_m B_m$ also lies in the estimation space.

**Note:**

If $By^T$ is the best estimator of a parametric function, then $B = \lambda A^T$. Also, if $Ly^T$ is an error function, then $LA = 0$. Therefore

$$BL^T = \lambda A^T L^T = 0,$$

showing that the best estimates of parametric functions are uncorrelated with the error functions. As we have seen the number of independent parametric functions that can be estimated is r, where r is the rank of A, and their best estimates are uncorrelated with the $(n - r)$ linear independent error functions.

## 2.3 Generalizations

The unique best unbiased estimator of $C\theta^T$ is obtained by substituting any solution of the weighted least square normal equations with the weight $\Gamma^{-1}$ where $y_1, \dots, y_n$ are n random variables with $E(y) = \theta A^T$ and the variance-covariance matrix of y is proportional to a known matrix $\Gamma$.

Let $E(y) = \theta A^T$ be the "observational equation" where $y = (y_1, \dots, y_n)$ and $\theta = (\theta_1, \dots, \theta_k)$ and $A = (a_{ij})$, $i = 1, \dots, n$, $j = 1, \dots, k$. Here $\theta_1, \dots, \theta_k$ are k unknown parameters and the coefficients $a_{ij}$ of the matrix A are known. Let the variance-covariance matrix of y be $\Gamma\sigma^2$ where $\sigma^2$ is an unknown constant.

Suppose we want to estimate the linear parametric function $C\theta^T$ where $C = (c_1, \dots, c_k)$. As we have already seen the necessary and sufficient condition for the estimability of $C\theta^T$ is that C should belong to the row space of A. Let us proceed to get the best-unbiased estimator in this case.

Let $By^T$ be a unbiased estimator of $C\theta^T$ so that $E(By^T) = C\theta^T$ and $BA = C$. Then the best unbiased estimator is obtained by minimizing $V(By^T)$. Now,

$$V(By^T) = B \cdot Cov(y^T, y) \cdot B^T$$
$$= B\Gamma B^T \cdot \sigma^2.$$

We have to minimize $B\Gamma B^T$ subject to the condition $BA = C$. Let $\lambda = (\lambda_1, \ldots, \lambda_k)$ be a Langrangian multiplier. Then, if

$$L = B\Gamma B^T - 2\lambda(BA - C)^T,$$

the minimizing equations are

$$\frac{\partial L}{\partial B^T} = B\Gamma - \lambda A^T = 0 \text{ and } \frac{\partial L}{\partial \lambda} = (BA - C)^T = 0.$$

$$\text{i.e., } B\Gamma = \lambda A^T \text{ and } BA = C.$$

The first of these equations gives $B = \lambda A^T \Gamma^{-1}$ and using the second we get

$$\lambda A^T \Gamma^{-1} A = C, \text{ giving } \lambda.$$

We know that the equations $BA = C$ are consistent and so we can always determine $\lambda$. Therefore $\lambda A^T \Gamma^{-1} y^T$ will always give an expression for $By^T$ in terms of y and this is an unbiased estimator with the minimum variance. The expression is also unique. For, if $\lambda_1$ and $\lambda_2$ are two values of $\lambda$ determined by $\lambda A^T \Gamma^{-1} A = C$, then

$$\lambda_1 A^T \Gamma^{-1} A = C, \ \lambda_2 A^T \Gamma^{-1} A = C$$

so that $(\lambda_1 - \lambda_2) A^T \Gamma^{-1} A = 0$. We have to prove that $\lambda_1 A^T \Gamma^{-1} y^T = \lambda_2 A^T \Gamma^{-1} y^T$. Let

$$P = (\lambda_1 - \lambda_2) A^T \Gamma^{-1} y^T.$$

Then it is easy to see that $E(P) = 0$ and $V(P) = 0$ so that $P = 0$. Thus $\lambda A^T \Gamma^{-1} y^T$ is unique. Thus $\lambda A^T \Gamma^{-1} y^T$ gives the unique best unbiased estimator of $C\theta^T$, where $\lambda$ is obtained from $\lambda A^T \Gamma^{-1} A = C$.

Next we show that this unique estimator is given by substituting any solution of the weighted least square normal equations in $C\theta^T$.

The weighted error sum of squares is

$$L = (y - \theta A^T)\Gamma^{-1}(y - \theta A^T)^T,$$

and this has to be minimized. The normal equations are given by

$$\frac{\partial L}{\partial \theta^T} = -2(y - \theta A^T)\Gamma^{-1}A = 0,$$

$$\text{i.e.,}\quad y\Gamma^{-1}A = \theta A^T\Gamma^{-1}A.$$

As before we can prove that these equations are consistent. Let $\hat{\theta} = (\hat{\theta}_1, \dots, \hat{\theta}_k)$ be any solution. Then $y\Gamma^{-1}A = \theta A^T\Gamma^{-1}A$. Substituting $\theta = \hat{\theta}$ in $C\theta^T$, we get

$$\begin{aligned} C(\hat{\theta})^T &= \lambda A^T\Gamma^{-1}A(\hat{\theta})^T \\ &= \lambda(\hat{\theta}A^T\Gamma^{-1}A)^T \\ &= \lambda(y\Gamma^{-1}A)^T \\ &= \lambda A^T\Gamma^{-1}y^T, \end{aligned}$$

which, as we have already shown, is the unique best unbiased estimator of $C\theta^T$.

Therefore, if $C\theta^T$ is estimable, the estimator is given by $C(\hat{\theta})^T$ where $\hat{\theta}$ satisfies the equation

$$y\Gamma^{-1}A = \hat{\theta}A^T\Gamma^{-1}A.$$

If $(A^T\Gamma^{-1}A)^{-1}$ exists then $\hat{\theta}$ is uniquely determined by $\hat{\theta} = y\Gamma^{-1}A(A^T\Gamma^{-1}A)^{-1}$ and then $C(\hat{\theta})^T$ gives the best possible estimator.

**Variance of the estimator**

The best unbiased estimator of $C\theta^T$ is given by $T = \lambda A^T\Gamma^{-1}y^T$ where $\lambda$ is given by

$$C = \lambda A^T\Gamma^{-1}A.$$

Therefore

$$\begin{aligned} V(T) &= V(\lambda A^T\Gamma^{-1}y^T) \\ &= \lambda A^T\Gamma^{-1}\mathrm{Cov}(y^T, y)\Gamma^{-1}A\lambda^T \end{aligned}$$

$$= \lambda A^T\Gamma^{-1}\Gamma\sigma^2\Gamma^{-1}A\lambda^T$$
$$= \lambda A^T\Gamma^{-1}A\lambda^T\sigma^2$$
$$= C\lambda^T\sigma^2.$$

In the particular case when $\theta$ is uniquely estimated by

$$\hat{\theta} = y\Gamma^{-1}A\left(A^T\Gamma^{-1}A\right)^{-1},$$

$$V(\hat{\theta}) = \left\{\left(A^T\Gamma^{-1}A\right)^{-1}\right\}^T A^T\Gamma^{-1}\mathrm{Cov}\left(y^T, y\right)\Gamma^{-1}A\left(A^T\Gamma^{-1}A\right)^{-1}$$
$$= \left(A^T\Gamma^{-1}A\right)^{-1}\sigma^2, \quad \text{since} \quad \mathrm{Cov}\left(y^T, y\right) = \Gamma\sigma^2.$$

Thus the diagonal elements of $\left(A^T\Gamma^{-1}A\right)^{-1}\sigma^2$ give the variances of $\hat{\theta}_1, \ldots, \hat{\theta}_k$, and the non-diagonal elements give the covariances between the $\hat{\theta}_i$'s.

If $C_1\theta^T$ and $C_2\theta^T$ are two estimable parametric functions, then $\lambda_1A^T\Gamma^{-1}y^T$ and $\lambda_2A^T\Gamma^{-1}y^T$ are respectively the unique best unbiased estimators where $\lambda_1$ and $\lambda_2$ are determined by the relations

$$\lambda_1A^T\Gamma^{-1}A = C_1 \text{ and } \lambda_2A^T\Gamma^{-1}A = C_2.$$

The covariance between the two estimators is

$$\mathrm{Cov}(\lambda_1A^T\Gamma^{-1}y^T, \lambda_2A^T\Gamma^{-1}y^T) = \lambda_1A^T\Gamma^{-1}\mathrm{Cov}\left(y^T, y\right)\Gamma^{-1}A{\lambda_2}^T$$
$$= \lambda_1A^T\Gamma^{-1}\mathrm{Cov}\left(y^T, y\right)\Gamma^{-1}A{\lambda_2}^T$$
$$= \lambda_1A^T\Gamma^{-1}A{\lambda_2}^T\sigma^2$$
$$= C_1{\lambda_2}^T\sigma^2 \text{ or } \lambda_1{C_2}^T\sigma^2.$$

**David-Neyman Generalization**

Suppose the variances of $y_1, \ldots, y_n$ are ${\sigma_1}^2, \ldots, {\sigma_n}^2$ respectively, where ${\sigma_1}^2, \ldots, {\sigma_n}^2$ are in given proportions, say, $p_1 : p_2 : \ldots : p_n$. (Here also the y's are assumed to be independent.). Thus

$$V(y_1) = p_1\sigma^2$$
$$V(y_2) = p_2\sigma^2$$
$$\dots\dots\dots$$
$$V(y_n) = p_n\sigma^2,$$

where $\sigma^2$ is an unknown constant. But then we have a special case of Rao's generalization with $\Gamma = \text{diag.}(p_1, \dots, p_n)$, since $\text{Cov}(y_i, y_j) = 0$.

## 2.4 Estimation of $\sigma^2$

Simple Case:

**Theorem 2.4**

Let r be the rank of A and $s^2$ be the minimum of $L = (y - \theta A^T)(y - \theta A^T)^T$ w.r.t. $\theta$. Then an unbiased estimator of $\sigma^2$ is given by $\frac{s^2}{(n-r)}$.

**Proof.**

Since the rank of A is r, A has a set of r linearly independent rows. From vector space theory we then know that there exist $(n - r)$ other mutually orthogonal (and therefore independent) rows each orthogonal to the rows of A. Further we may take these $(n - r)$ vectors as normalized also. Consider now the $(n - r) \times n$ matrix D formed by these $(n - r)$ vectors. Then

$$A^T D = 0, \text{ and } D^T D = I$$

The rows of A and D together contain a set of n linearly independent vectors and as such provide a basis of the n-vector space. Any n-vector can therefore be represented as a linear combination of these two sets of vectors. In particular y can be written as

$$y = \alpha A^T + \beta D^T$$

where $\alpha = (\alpha_1, \ldots, \alpha_k)$ and $\beta = (\beta_1, \ldots, \beta_{n-r})$.

With this representation, we have

$$\begin{aligned}
L &= (y - \theta A^T)(y - \theta A^T)^T \\
&= (\alpha A^T + \beta D^T - \theta A^T)(\alpha A^T + \beta D^T - \theta A^T)^T \\
&= [(\alpha - \theta)A^T + \beta D^T][(\alpha - \theta)A^T + \beta D^T]^T \\
&= (\alpha - \theta)A^T A(\alpha - \theta)^T + (\alpha - \theta)A^T D\beta^T + \beta D^T A(\alpha - \theta)^T + \beta D^T D\beta^T \\
&= (\alpha - \theta)A^T A(\alpha - \theta)^T + \beta\beta^T.
\end{aligned}$$

We have to minimize $L$ w.r.t. $\theta$. Obviously, $\beta\beta^T \geq 0$. Also, $(\alpha - \theta)A^T A(\alpha - \theta)$ is a quadratic form with matrix $A^T A$. So it is at least positive semi-definite. Therefore its minimum value is zero, which is the case when $\alpha = \theta$. Therefore, if the minimum of L be denoted by $L_{min}$, then $L_{min} = \beta\beta^T = s^2$, according to our notation. Now, since $y = \alpha A^T + \beta D^T$, we have

$$yD = \alpha A^T D + \beta D^T D = \beta, \text{ since } A^T D = 0.$$

So,

$$E(\beta) = E(yD) = E(y)D = \theta A^T D = 0,$$

and

$$\begin{aligned}
V(\beta) = V(yD) &= D^T \cdot \mathrm{Cov}(y^T, y) \cdot D \\
&= D^T D\sigma^2 \\
&= I\sigma^2.
\end{aligned}$$

Thus each $\beta_i$ has mean zero and variance $\sigma^2$. Hence

$$E(s^2) = E(\beta\beta^T)$$
$$= E\left(\sum_1^{n-r} \beta_i^{\,2}\right)$$
$$= \sum_1^{n-r} E(\beta_i^{\,2})$$
$$= (n-r)\sigma^2.$$

So $E\left(\frac{s^2}{n-r}\right) = E\left(\frac{L_{min}}{n-r}\right) = \sigma^2$. Showing that $\frac{s^2}{n-r} = \frac{L_{min}}{n-r}$ is an unbiased estimator of $\sigma^2$. ∎

**Note 1.**

We have seen that the variance of the best unbiased estimator of $C\theta^T$ is $C\lambda^T\sigma^2$ and so an unbiased estimator of this variance will be $C\lambda^T\frac{s^2}{n-r}$.

**Note 2.**

The relation $\beta = yD$ shows that if the y's are normally distributed, then the $\beta$'s being linear functions of the y's are also normally distributed. Further $E(\beta_i) = 0, V(\beta_i) = \sigma^2$ and $Cov(\beta_i, \beta_j) = 0$. Hence $\beta_1, \dots, \beta_{n-r}$ are independent normal variates with mean zero and variance $\sigma^2$. Therefore

$$\frac{s^2}{\sigma^2} = \sum_1^{n-r}\left(\frac{\beta_i}{\sigma}\right)^2$$

has a chi-square distribution with $(n-r)$ d.f. (In asserting the independence of the $\beta$'s we used the fact that in the case of normally distributed variates non-correlation is equivalent to independence.) $s^2$ is called the unconditional minimum.

**General Case**

Here also we can prove that an unbiased estimator of $\sigma^2$ is given by $\frac{L_{min}}{n-r}$, where $L = (y - \theta A^T)\Gamma^{-1}(y - \theta A^T)^T$, and $\Gamma$ is proportional to the variance-covariance matrix of the y's. In this case D is the matrix obtained by normalizing the $(n - r)$ linearly independent vectors in the error space. But here D is orthogonal not only to A but also to $A\Gamma^{-1}$ and satisfies $D^T\Gamma^{-1}D = I$. As before, let

$$y = \alpha A^T + \beta D^T$$

so that $y\Gamma^{-1}D = \alpha A^T\Gamma^{-1}D + \beta\, D^T\Gamma^{-1}D = \beta$, since $A^T\Gamma^{-1}D = 0$ and $D^T\Gamma^{-1}D = I$. Thus

$$\begin{aligned} E(\beta) &= E(y\Gamma^{-1}D) \\ &= \theta A^T\Gamma^{-1}D \\ &= 0 \end{aligned}$$

and

$$\begin{aligned} V(\beta) &= V(y\Gamma^{-1}D) \\ &= D^T\Gamma^{-1}\mathrm{Cov}(y^T, y)\Gamma^{-1}D \\ &= D^T\Gamma^{-1}\Gamma\sigma^2\Gamma^{-1}D \\ &= D^T\Gamma^{-1}D\sigma^2 \\ &= I\sigma^2 \end{aligned}$$

Thus for each i, $E(\beta_i) = 0$, and $V(\beta_i) = \sigma^2$. Further $\mathrm{Cov}(\beta_i, \beta_j) = 0$. Now,

$$\begin{aligned} L &= (y - \theta A^T)\Gamma^{-1}(y - \theta A^T)^T \\ &= (\alpha A^T + \beta\, D^T - \theta A^T)\Gamma^{-1}(\alpha A^T + \beta\, D^T - \theta A^T)^T \\ &= [(\alpha - \theta)A^T + \beta\, D^T]\Gamma^{-1}[(\alpha - \theta)A^T + \beta\, D^T]^T \\ &= (\alpha - \theta)A^T\Gamma^{-1}A(\alpha - \theta) + \beta\beta^T. \end{aligned}$$

and by a similar argument as in the previous case it can be seen that

$$L_{min} = \beta\beta^T = \sum_{i=1}^{n-r} \beta_i = s^2.$$

Therefore $E(s^2) = \sum_{i=1}^{n-r} E(\beta_i{}^2) = (n-r)\sigma^2$ or $E\left(\frac{s^2}{n-r}\right) = \sigma^2$.

Thus $\frac{s^2}{n-r} = \frac{L_{min}}{n-r}$ is an unbiased estimator of $\sigma^2$.

In this case also if the y's are normally distributed, then $\frac{s^2}{n-r}$ will be distributed as a $\chi^2$ with $(n-r)$ d.f.. Note that

$$\begin{aligned} n-r &= (\text{total number of variables}) - (\text{rank of A}) \\ &= (\text{total nubmer of variables}) \\ &\qquad - (\text{the total number of independent} \\ &\qquad \text{linear estimable parametric functions.}) \end{aligned}$$

## 2.5 Observational Equations with Linear Restrictions On The Parameters

Sometimes it may be known that the parameters $\theta_1, \dots, \theta_k$ in the observational equations satisfy some linear restrictions of the form

$$g_i = p_{i1}\theta_1 + \cdots + p_{ik}\theta_k\,; \quad i = 1,2, \dots, s. \qquad (1)$$

In this situation two courses are open. It may be possible to eliminate some of the parameters in the observational equations with the help of equations (1) and obtain a different set of observational equations with fewer parameters having no restrictions. The theory developed above will then be applicable.

Another method is to derive the normal equations by minimizing the error sum of squares, subject to the restrictions (i). In matrix notation, the restrictions on the parameters become

$$\theta P^T = G^T$$

where $P = \begin{pmatrix} p_{11} & \cdots & p_{1k} \\ \cdots & \cdots & \cdots \\ p_{s1} & \cdots & p_{sk} \end{pmatrix}$ and $G = \begin{pmatrix} g_1 \\ \vdots \\ g_s \end{pmatrix}$.

Let t be the number of independent row vectors of P which can be represented as linear combinations of the rows of A. (If there is no such vector in P, then t = 0). Then, without loss of generality, we may write P as

$$P = \begin{pmatrix} P_1 \\ P_2 \end{pmatrix}$$

where $P_1$ contains the t vectors of P which can be represented as linear combinations of the rows of A and $P_2$ contains the remaining $s - t$ vectors. Let the corresponding partition of G be

$$G = \begin{pmatrix} G_1 \\ G_2 \end{pmatrix}.$$

Then the restrictions on the θ's become

$$\theta {P_1}^T = {G_1}^T \text{ and } \theta {P_2}^T = {G_2}^T.$$

Let D be an orthogonal matrix consisting of a set of $(n - r)$ linear independent column vectors which are orthogonal to the vectors of A. Then $D^T D = I$ and $D^T A = 0$. The r independent vectors in A together with the $(n - r)$ independent vector in D form a basis of the n-vector space. Therefore, in general, any n-vector can be represented as a linear combination of the vectors of A and D. Thus we may want

$$y = \alpha A^T + \beta D^T$$

where $\alpha = (\alpha_1, \dots, \alpha_k)$ and $\beta = (\beta_1, \dots, \beta_{n-r})$.

We have to minimize $(y - \theta A^T)(y - \theta A^T)^T$ subject to the condition $\theta P^T = G^T$. Introduce the Langrangian multipliers $l_1, \dots, l_s$. Let

$$\lambda = (l_1, \dots, l_s).$$

Write $\lambda = (\lambda_1 \;\; \lambda_2)$ where $\lambda_1$ contains the first t multipliers and $\lambda_2$ contains the last $(s - t)$ multipliers. Consider the function

$$L = (y - \theta A^T)(y - \theta A^T)^T + 2\lambda_1(P_1\theta^T - G_1) + 2\lambda_2(P_2\theta^T - G_2)$$

Replacing y by $\alpha A^T + \beta D^T$, we get

$$L = (\alpha - \theta)A^T A(\alpha - \theta)^T + \beta\beta^T + 2\lambda_1(P_1\theta^T - G_1) + 2\lambda_2(P_2\theta^T - G_2).$$

Differentiating w.r.t. $\theta, \lambda_1, \lambda_2$ and equating to zero, the minimizing equations are

$$-2(\alpha - \theta)A^T A + 2\lambda_1 P_1 + 2\lambda_2 P_2 = 0,$$

$$P_1\theta^T - G_1 = 0, \quad P_2\theta^T - G_2 = 0.$$

$$\text{i.e.,} \quad (\alpha - \theta)A^T A - \lambda_1 P_1 - \lambda_2 P_2 = 0 \qquad (2)$$

$$P_1\theta^T = G_1 \qquad (3)$$

$$P_2\theta^T = G_2. \qquad (4)$$

These are the normal equations which are $(s + k)$ in number. From (2), we get

$$(\alpha - \theta)A^T A - \lambda_1 P_1 = \lambda_2 P_2.$$

The L. H. S. is a linear combination of the vectors of A and the R. H. S. is independent of the vectors of A. So $\lambda_2 P_2 = 0$. But at this stage we cannot assume that $P_2 = 0$, since we do not know whether all vectors of P can be represented as linear combinations of the vectors of A, and hence the only alternative is $\lambda_2 = 0$. But then (2) becomes

$$(\alpha - \theta)A^T A = \lambda_1 P_1 \qquad (5).$$

Now, $yA = \alpha A^T A + \beta D^T A = \alpha A^T A$. Let T be such that $TA^T A = P_1$. Then

$$\alpha P_1{}^T = \alpha A^T A T^T = yAT^T,$$

so that $E(\alpha P_1{}^T) = E(yAT^T) = \theta A^T A T^T = \theta P_1{}^T$. Hence

$$E(\alpha P_1{}^T - \theta P_1{}^T) = 0.$$

Let $Z = (\alpha - \theta)P_1{}^T$. Then $E(Z) = 0$, and

$$
\begin{aligned}
V(Z) &= V(\alpha P_1{}^T - \theta P_1{}^T) \\
&= V(yAT^T - \theta P_1{}^T) \\
&= V(yAT^T), \text{ since } \theta P_1{}^T \text{ is a constant.} \\
&= TA^T \mathrm{Cov}(y^T, y) AT^T \\
&= TA^T AT^T \sigma^2 \\
&= P_1 T^T \sigma^2 \\
&= \Gamma\sigma^2, \text{ say, where } \Gamma = P_1 T^T.
\end{aligned}
$$

But $\Gamma\sigma^2$, being a variance-covariance matrix, is symmetric and non-singular. So $\Gamma^{-1}$ exists.

Consider the equation (5).

$$(\alpha - \theta)A^T A = \lambda_1 P_1.$$

Post-multiplying both sides by $T^T$, we get

$$(\alpha - \theta)A^T AT^T = \lambda_1 P_1 T^T$$

$$\text{i.e., } (\alpha - \theta)P_1{}^T = \lambda_1 \Gamma, \text{ or } \lambda_1 = (\alpha - \theta)P_1{}^T \Gamma^{-1}.$$

Let ${s_0}^2$ be the minimum of L. then

$${s_0}^2 = L_{min} = (\alpha - \theta)A^T A(\alpha - \theta)^T + \beta\beta^T$$

where

$$
\begin{aligned}
(\alpha - \theta)A^T A &= \lambda_1 P_1 \\
&= \lambda_1 P_1 (\alpha - \theta)^T + \beta\beta^T \\
&= (\alpha - \theta)P_1{}^T \Gamma^{-1} P_1 (\alpha - \theta)^T + \beta\beta^T \\
&= Z\Gamma^{-1} Z^T + \beta\beta^T \\
&= Z\Gamma^{-1} Z^T + s^2,
\end{aligned}
$$

where $s^2$ is the unconditional minimum.

If $y_1, \dots, y_n$ are independently and normally distributed with a common variance $\sigma^2$, then $\frac{s^2}{\sigma^2}$ is distributed as a $\chi^2$ with $(n-r)$ d.f.. Now Z is linear combination of the y's with $E(Z) = 0$ and $V(Z) = \Gamma\sigma^2$, and therefore $\frac{Z\Gamma^{-1}Z^T}{\sigma^2}$ is also distributed as a $\chi^2$ with t d.f., (There are t independent vectors in Z). Hence, if $s^2$ and $Z\Gamma^{-1}Z^T$ are independent, then $\frac{s_0{}^2}{\sigma^2}$ is distributed as a $\chi^2$ with $(n-r+t)$ d.f..

Next, we shall show that they are independently distributed. We have $\beta = yD$ and $Z = (\alpha - \theta)P_1{}^T = yAA^T - \theta P_1{}^T$, since $\alpha P_1{}^T = yAT^T$. Therefore

$$\begin{aligned}
\mathrm{Cov}(Z, \beta) &= \mathrm{Cov}(yD, yAT^T - \theta P_1{}^T) \\
&= \mathrm{Cov}(yD, yAT^T), \text{ since } \theta P_1{}^T \text{ is constant} \\
&= D^T\mathrm{Cov}(y^T, y)AT^T \\
&= D^TAT^T\sigma^2 \\
&= 0.
\end{aligned}$$

Thus Z and $\beta$ are uncorrelated. Further they are normally distributed. Hence Z and $\beta$ are independent. So $Z\Gamma^{-1}Z^T$ and $\beta\beta^T = s^2$ and independent.

Again, $\frac{s_0{}^2 - s^2}{\sigma^2}$ is a $\chi^2$ with t d.f. and so $\frac{(s_0{}^2 - s^2)/t}{s^2/(n-r)}$ is an F with t and $(n-r)$ d.f. (t is the total number of independent rows of P which could be obtained as linear combination of the rows of A.)

**Tests of Linear Hypotheses**

A linear hypothesis of the form

$$H_0\colon\ \theta P^T = G^T$$

where P is an $s \times m$ matrix $(s \le k)$, specifies the values of one or more linear functions of the parameters. If the rank of P is t, $H_0$ will be called a hypothesis of rank t. The above results provide a criterion to test the hypothesis $H_0$ when $y_1, \dots, y_n$ are normally distributed with a common variance $\sigma^2$. We have

$$\frac{({s_0}^2 - s^2)/t}{s^2/n - r} = F(t, n - r).$$

This is a test for $H_0$ because the numerator will be a $\chi^2$ if and only if $H_0$ is true. If F is significant at a certain level, then the θ's are not in agreement with the restrictions imposed. But if F is not significant at the level, we may impose the restrictions.

## Applications

### I. Linear Regression

Let $(x_1, y_1), \ldots, (x_n, y_n)$ be a sample of n observations on a bivariate random variable $(x, y)$. If we assume that y is linearly dependent on the x, then the regression model may be written as

$$E(y) = a + bx$$

where a and b are parameters.
The observational equations then become

$$E(y) = \theta A^T$$

where $y = (y_1, \ldots, y_n)$, $\theta = (a, b)$ and $A = \begin{pmatrix} 1 & x_1 \\ 1 & x_2 \\ \vdots & \vdots \\ 1 & x_n \end{pmatrix}$. After simplification the normal equations become

$$\sum y = na + b\sum x$$

$$\sum xy = a\sum x + b\sum x^2.$$

and the estimates for a and b work out to be

$$\hat{a} = \bar{y} - \hat{b}\bar{x}, \quad \hat{b} = \frac{\frac{1}{n}\sum xy - \bar{x}\bar{y}}{\frac{1}{n}\sum x^2 - \bar{x}^2}.$$

Now,

$$
\begin{aligned}
s^2 &= L_{min} \\
&= (y - \hat{\theta}A^T)y^T \\
&= yy^T - \hat{\theta}A^T y^T \\
&= \sum y^2 - (\hat{a} \quad \hat{b})\begin{pmatrix} 1 & 1 & \cdots & 1 \\ x_1 & x_2 & \cdots & x_n \end{pmatrix}\begin{pmatrix} y_1 \\ \vdots \\ y_n \end{pmatrix} \\
&= \sum y^2 - \hat{a}\sum y - \hat{b}\sum xy \\
&= \sum y^2 - (\bar{y} - \hat{b}\bar{x})\sum y - \hat{b}\sum xy \\
&= n[\frac{1}{n}\sum y^2 - \bar{y}^2 - \hat{b}(\frac{1}{n}\sum xy - \bar{x}\bar{y})] \\
&= n\left[\frac{1}{n}\sum y^2 - \bar{y}^2 - \frac{\left(\frac{1}{n}\Sigma xy - \bar{x}\bar{y}\right)^2}{\frac{1}{n}\Sigma x^2 - \bar{x}^2}\right] \\
&= n[s_y^2 - \frac{(\mathrm{Cov}(x, y))^2}{s_x^2}]
\end{aligned}
$$

where $s_x^2 = \frac{1}{n}\Sigma x^2 - \bar{x}^2$, $s_y^2 = \Sigma y^2 - \bar{y}^2$, and

$$
\begin{aligned}
\mathrm{Cov}(x, y) &= \frac{1}{n}\sum xy - \bar{x}\bar{y} \\
&= s_{xy} \\
&= ns_y^2(1 - r_{xy}^2)
\end{aligned}
$$

where $r_{xy}$ is the sample correlation between x and y.

**Test of hypothesis that there is no regression**

$H_0\colon b = 0$. This is a hypothesis with one d. f.. The minimizing function after imposing the condition is

$$L = \sum (y - a)^2$$

and the only normal equation in this case is

$$\sum (y - a) = 0$$

which estimates a to be $\hat{a} = \bar{y}$. Hence the conditional minimum is

$$s_0^2 = \sum (y - \bar{y})^2 = ns_y^2,$$

which has $(n - 1)$ d. f.. Therefore

$$s_0^2 - s^2 = ns_y^2 r_{xy}^2$$

with 1 d. f., and $s^2$ has $(n - 2)$ d. f.. Assuming normal distribution for the y's , the F-ratio is given by

$$F(1, n - 2) = (n - 2)\frac{r_{xy}^2}{1 - r_{xy}^2}.$$

**To test $H_0$: $b = b_0$**

This is a linear hypothesis with 1 d. f.. Assuming $H_0$, the regression model is

$$E(y) = a + b_0 x.$$

The conditional minimum works out to be

$$s_0^2 = \sum (y - \hat{a} - b_0 x)^2 , \hat{a} = \bar{y} - b_0 \bar{x}.$$

$$= \sum (y - \bar{y})^2 + b_0^2 (x - \bar{x})^2 - 2b_0 \sum (x - \bar{x})(y - \bar{y}).$$

After simplifications, we get

$$s_0^2 - s^2 = n\, s_x^2 (\hat{b} - b_0)^2,$$

where $\hat{b} = \frac{\mathrm{Cov}(x,y)}{s_x^2}$, the estimate of b under no condition. Thus, in this case, the test criterion is

$$F(1, n-2) = \frac{\dfrac{s_x^2(\hat{b} - b_0)^2}{1}}{\dfrac{s_y^2(1 - r_{xy}^2)}{n-2}}.$$

**t-test**

Here we can apply the t-test also. When the parameters are estimable (here a and b are uniquely estimable) we have seen that $V(\hat{\theta}) = (A^T A)^{-1}\sigma^2$ where $\sigma^2$ is the unknown population variance. Here $\theta = (a, b)$ and

$$A = \begin{pmatrix} 1 & x_1 \\ 1 & x_2 \\ \cdots & \cdots \\ 1 & x_n \end{pmatrix},$$

$$A^T A = \begin{pmatrix} n & \sum x \\ \sum x & \sum x^2 \end{pmatrix}$$

and

$$(A^T A)^{-1} = \frac{1}{n\sum x^2 - (\sum x)^2} \begin{pmatrix} \sum x^2 & -\sum x \\ -\sum x & n \end{pmatrix}.$$

Hence

$$V(\hat{a}) = \frac{\sum x^2}{n\sum x^2 - (\sum x)^2}\sigma^2, \text{ and } V(\hat{b}) = \frac{n}{n\sum x^2 - (\sum x)^2}\sigma^2.$$

An unbiased estimator of $\sigma^2$ is given by

$$\frac{L_{min}}{n-r} = \frac{s^2}{n-2}, \text{ where, in this case, } s^2 = ns_y^2(1 - r_{xy}^2).$$

Under the assumption that the y's are normally and independently distributed, the test criterion is

$$t_{n-2} = \frac{\hat{b} - b_0}{\sqrt{\text{An unbiased estimate of } V(\hat{b})}}$$

$$= \frac{\hat{b} - b_0}{\sqrt{\frac{n}{n\sum x^2 - (\sum x)^2} \frac{s^2}{n-2}}}.$$

**Generalization**

Suppose we draw k samples of sizes $n_1, \ldots, n_k$ from a bivariate population. Let the data be as given below

$$1. \begin{cases} y_{11} & y_{12} & \cdots & y_{1n_1} \\ x_{11} & x_{12} & \cdots & x_{1n_1} \end{cases}$$

$$2. \begin{cases} y_{21} & y_{22} & \cdots & y_{2n_2} \\ x_{21} & x_{22} & \cdots & x_{2n_2} \end{cases}$$

… … … … … … … …

$$k. \begin{cases} y_{k1} & y_{k2} & \cdots & y_{kn_k} \\ x_{k1} & x_{k2} & \cdots & x_{kn_k} \end{cases}$$

Let the regression model for the i-th sample be

$$E(y_{ij}) = a_i + b_i x_{ij}, j = 1, \ldots, n_i .$$

The error sum of squares for the i-th sample is

$$\sum_{j=1}^{n_i} (y_{ij} - a_i - b_i x_{ij})^2.$$

So the total error sum of squares is

$$L = \sum_{i,j} (y_{ij} - a_i - b_i x_{ij})^2.$$

The minimizing equations are

$$\frac{\partial L}{\partial a_i} = 0 \text{ and } \frac{\partial L}{\partial b_i} = 0, i = 1, \ldots, k$$

$$\text{i.e., } \sum_j (y_{ij} - a_i - b_i x_{ij}) = 0 \quad (6)$$

$$\sum_j (y_{ij} - a_i - b_i x_{ij})\, x_{ij} = 0 \quad (7)$$

These are the normal equations which are 2k in number. Solving them we get estimates for the 2k parameters $a_1, \ldots, a_k;\ b_1, \ldots, b_k$.

From equation (6), we get

$$\sum_j y_{ij} - \sum_j a_i - \sum_j b_i x_{ij} = 0$$

$$\text{i.e., } y_{i\cdot} - n_i a_i - b_i x_{i\cdot} = 0,$$

where $y_{i\cdot} = \sum_j y_{ij}$ and $x_{i\cdot} = \sum_j x_{ij}$. This gives

$$\hat{a}_i = \frac{1}{n_i}(y_{i\cdot} - \hat{b}_i x_{i\cdot}) = \overline{y_{i\cdot}} - \hat{b}_i \bar{x}_{i\cdot}.$$

From (7), we get

$$\sum_j x_{ij} y_{ij} - \hat{a}_i \sum_j x_{ij} - \hat{b}_i \sum_j x_{ij}^2 = 0.$$

$$\text{i.e., } \sum_j x_{ij} y_{ij} - (\bar{y}_{i\cdot} - \hat{b}_i \bar{x}_{i\cdot}) \sum_j x_{ij} - \hat{b}_i \sum_j x_{ij}^2 = 0.$$

$$\text{i.e., } \sum_j x_{ij} y_{ij} - \bar{y}_{i\cdot} \sum_j x_{ij} = \hat{b}_i (\sum_j x_{ij}^2 - \bar{x}_{i\cdot} \sum_j x_{ij})$$

$$\text{i.e., } \sum_j x_{ij} y_{ij} - n_i \bar{y}_{i\cdot} \bar{x}_{i\cdot} = \hat{b}_i \left( \sum_j x_{ij}^2 - n_i \bar{x}_{i\cdot}^2 \right)$$

Put $A_i = \sum_j y_{ij}^2 - n_i \bar{y}_{i\cdot}^2$, $B_i = \sum_j x_{ij}^2 - n_i \bar{x}_{i\cdot}^2$ and $C_i = \sum_j x_{ij} y_{ij} - n_i \bar{y}_{i\cdot} \bar{x}_{i\cdot}$. With these notations, $\hat{b}_i = \frac{C_i}{B_i}$.

Therefore the unconditional minimum is

$$s^2 = \text{minimim of} \sum_{i,j} (y_{ij} - a_i - b_i x_{ij})^2$$

$$= \sum_{i,j} (y_{ij} - \hat{a}_i - \hat{b}_i x_{ij})\, y_{ij},$$

since the other terms vanish because of the normal equations. Therefore

$$s^2 = \sum_{i,j} (y_{ij} - \bar{y}_{i\cdot} + \hat{b}_i \bar{x}_{i\cdot} - \hat{b}_i x_{ij})\, y_{ij}$$

$$= \sum_{i,j} (y_{ij}^2 - \bar{y}_{i\cdot} y_{ij}) - \sum_{i,j} \hat{b}_i (x_{ij} y_{ij} - \bar{x}_{i\cdot} y_{ij})$$

$$= \sum_i \left( \sum_j y_{ij}^2 - n_i \bar{y}_{i\cdot}^2 \right) - \sum_i \hat{b}_i \left( \sum_j x_{ij} y_{ij} - n_i \bar{x}_{i\cdot} \bar{y}_{i\cdot} \right)$$

$$= \sum_i A_i - \sum_i \hat{b}_i C_i$$

$$= \sum_i A_i - \sum_i \frac{C_i^2}{B_i}$$

$$= \sum_i (A_i - \frac{C_i^2}{B_i}).$$

This has $n. - 2k$ degrees of freedom where $n. = \sum_i n_i$.

**To test $H_0$: $b_1 = b_2 = \cdots = b_k$.**

This is a linear hypothesis with $(k-1)$ d.f. and it enables us to test whether the rate of change of y w.r.t. x is the same in all the samples.

Let $b_1 = b_2 = \cdots = b_k = b$, when $H_0$ is true. The least square estimates are

$$\hat{a}_i = \bar{y}_{i\cdot} - \hat{b}\bar{x}_{i\cdot}, \quad \hat{b} = \frac{\sum_i C_i}{\sum_i B_i}.$$

The conditional minimum is

$$s_0^2 = \sum_{i,j} (y_{ij} - \hat{a}_i - \hat{b}x_{ij})\, y_{ij}$$

$$= \sum_i \left( A_i - C_i \frac{\sum_i C_i}{\sum_i B_i} \right), \quad \text{on simplification.}$$

This has $n. -2k + (k-1) = n. -k - 1$ d.f.. So under the assumption that the y's are independent normal variates with common variance $\sigma^2$, $\frac{s^2}{\sigma^2}$ is a $\chi^2$ with $(n. -k - 1)$ d.f. The F-ratio for testing $H_0$ is

$$F(k-1, n. -2k) = \frac{(s_0^2 - s^2)/(k-1)}{s^2/(n. -2k)}$$

**t-test**

Here $\sigma^2$ is unknown and so if we can construct unbiased estimators for the variances of the estimators of the parameters then we may make use of the t-test to test individual hypotheses. The observational equations are

$$E(y_{ij}) = a_i + b_i x_{ij} \quad j = 1, \ldots, n_i; \quad i = 1, \ldots, k$$

Let $y = (y_{11}, \ldots, y_{1n_1}, y_{21}, \ldots, y_{2n_2}, \ldots, y_{k1}, \ldots, y_{kn_k})$ and $\theta = (a_1, \ldots, a_k, b_1, \ldots, b_k)$. Then all the observational equations telescope into the single matrix equation

$$E(y) = \theta A^T,$$

where $A^T$

$$= \begin{bmatrix} 1 & 1 & \dots & 1 & 0 & 0 & \dots & 0 & \dots & 0 & 0 & \dots & 0 \\ 0 & 0 & \dots & 0 & 1 & 1 & \dots & 1 & \dots & 0 & 0 & \dots & 0 \\ \dots & \dots & \dots & \dots & \dots & \dots & \dots & \dots & \dots & \dots & \dots & \dots & \dots \\ 0 & 0 & \dots & 0 & 0 & 0 & \dots & 0 & \dots & 1 & 1 & \dots & 1 \\ x_{11} & x_{12} & \dots & x_{1n_1} & 0 & 0 & \dots & 0 & \dots & 0 & 0 & \dots & 0 \\ 0 & 0 & \dots & 0 & x_{21} & x_{22} & \dots & x_{2n_2} & \dots & 0 & 0 & \dots & 0 \\ \dots & \dots & \dots & \dots & \dots & \dots & \dots & \dots & \dots & \dots & \dots & \dots & \dots \\ 0 & 0 & \dots & 0 & 0 & 0 & \dots & 0 & \dots & x_{k1} & x_{k2} & \dots & x_{kn_k} \end{bmatrix}$$

Thus A is an n.× 2k matrix and none of its rows is a linear combination of the rest, so that its rank is 2k. Consequently all the 2k parameters are uniquely estimable in which case,

$$V(\hat{\theta}) = \left(A^T A\right)^{-1} \sigma^2,$$

where $\sigma^2$ is the unknown population variance. We have,

$$A^T A = \begin{bmatrix} n_1 & 0 & \dots & 0 & \sum_j x_{1j} & 0 & \dots & 0 \\ 0 & n_2 & \dots & 0 & 0 & \sum_j x_{2j} & \dots & 0 \\ \dots & \dots & \dots & \dots & \dots & \dots & \dots & \dots \\ 0 & 0 & \dots & n_k & 0 & 0 & \dots & \sum_j x_{kj} \\ \sum_j x_{1j} & 0 & \dots & 0 & \sum_j x_{1j}^2 & 0 & \dots & 0 \\ 0 & \sum_j x_{2j} & \dots & 0 & 0 & \sum_j x_{2j}^2 & \dots & 0 \\ \dots & \dots & \dots & \dots & \dots & \dots & \dots & \dots \\ 0 & 0 & \dots & \sum_j x_{kj} & 0 & 0 & \dots & \sum_j x_{kj}^2 \end{bmatrix}$$

Let $P = \left(A^T A\right)^{-1}$. Then

$$V(\hat{a}_i) = p_{ii}\sigma^2; i = 1, \ldots, k \quad \text{and} \quad V(\hat{b}_i) = p_{ii}\sigma^2; i = k+1, \ldots, 2k$$

An unbiased estimate of $\sigma^2$ is $\frac{s^2}{n.-2k}$ where $s^2$ is the conditional minimum. Substituting this value in the above expressions we get unbiased estimates of the variances of estimators of all the parameters.

If we want to test the hypothesis $b_i = b_j$, the statistic to be constructed is

$$t = \frac{\hat{b}_i - \hat{b}_j}{\sqrt{\text{An unbiased estimate of } V(\hat{b}_i - \hat{b}_j)}}.$$

**Note:**

Suppose the F-ratio for testing $H_0: b_1 = b_2 = \cdots = b_k$ is significant. Based on this, we cannot reject the hypothesis $b_i = b_j$ or any other sub-hypothesis. We will have to develop individual tests for each of these hypotheses and the most appropriate test in this case is the t-test. Therefore if F is found significant, we proceed to apply the t-test to test individual hypotheses. But if F is not significant we are at full liberty to assume individual of sub-hypotheses such as $b_i = b_j$, $b_i = b_j = b_k$, etc.

One additional point is also worth mentioning here. In testing the individual hypotheses $b_i = b_j$ and $b_j = b_k$ suppose we find that the corresponding t's are not significant at a certain level. But from these, in general, we cannot conclude that the difference between $b_i$ and $b_k$ is not significant. In particular cases this may be true, but a further test will help us to draw the correct conclusion.

**To test $\mathbf{H_0}$**: $a_1 = \cdots = a_k;\ b_1 = \cdots = b_k$.

This is a linear hypothesis with $(k-1) + (k-1) = 2(k-1)$ d.f. Let under $H_0$,

$$a_1 = \cdots = a_k = a \quad \text{and} \quad b_1 \ldots = b_k = b.$$

Then, the total error sum of squares is

$$L = \sum_{i,j} (y_{ij} - a - bx_{ij})^2.$$

The minimizing equations are $\frac{\partial L}{\partial a} = 0$, $\frac{\partial L}{\partial b} = 0$, which give the normal equations

$$\Sigma_{i,j}(y_{ij} - a - bx_{ij}) = 0$$

$$\text{and } \Sigma_{i,j}(y_{ij} - a - bx_{ij})x_{ij} = 0$$

The least square estimates are

$$\hat{a} = \bar{y}_{..} - \hat{b}\bar{x}_{..} \text{ and } \hat{b} = \frac{C}{B},$$

where

$$\bar{y}_{..} = \frac{1}{n.}\sum_{i,j} y_{ij}\ ,\ \bar{x}_{..} = \frac{1}{n.}\sum_{i,j} x_{ij}\ ,\ C = \sum_{i,j} x_{ij}\, y_{ij} - n.\bar{x}_{..}\bar{y}_{..}\ ,\ B$$

$$= \sum_{i,j} x_{ij}^2 - n.\bar{x}_{..}^2\ .$$

Therefore the conditional minimum is

$$s_0^2 = \sum_{i,j} (y_{ij} - \hat{a} - \hat{b}x_{ij})\, y_{ij}$$

$$= A - \hat{b}C$$

$$= A - \frac{C^2}{B}$$

where

$$A = \sum_{i,j} y_{ij}^2 - n.\bar{y}_{..}^2\ .$$

Under the assumption of normality, $\frac{s_0^2}{\sigma^2}$ is a $\chi^2$ with $n.-2k+2(k-1)=$ $n.-2$ d.f. and $\frac{s^2}{\sigma^2}$ is a $\chi^2$ with $n.-2k$ d.f. where $s^2=\sum_i(A_i-\frac{C_i^2}{B_i})$.
The F-ratio is give by

$$F(2k-2,n.-2k)=\frac{(s_0^2-s^2)/2(k-1)}{s^2/(n.-2k)}.$$

**II. Multiple Regression**

Let the observations y be expressible as linear functions of some known variables $x_1,\dots,x_k$ with residual errors which are normally and independently distributed around zero with a constant variance $\sigma^2$. The regression model is then

$$E(y)=b_1^1x_1+\cdots+b_k^1x_k.$$

Let us change the origin to the mean and let the new regression model be

$$E(y)=b_1x_1+\cdots+b_kx_k.$$

Suppose we have a sample of n observations. The total error sum of squares is

$$L=\sum_i(y_i-b_1x_{1i}-\cdots-b_kx_{ki})^2.$$

The least square normal equations are given by

$$\frac{\partial L}{\partial b_j}=0, j=1,2,\dots,k.$$

$$\text{i.e.,}\quad \sum_i x_{ji}(y_i-b_1x_{1i}-\cdots-b_kx_{ki})=0;\quad j=1,\dots,k.$$

$$\text{i.e.,}\quad \sum_i x_{ji}y_i=b_1\sum_i x_{1i}x_{ji}+\cdots+b_k\sum_i x_{ki}x_{ji}.$$

Let $\sum_i x_{ji} y_i = Q_j$, $j = 1, \dots, k$ and $\sum_i x_{li} x_{ji} = s_{lj}$, $l = 1, \dots, k$. Then the normal equations become

$$Q_j = b_1 s_{1j} + \cdots + b_k s_{kj}; \; j = 1, 2, \dots, k$$

In matrix notation, these become $Q = BS$, where $Q = (Q_1, \dots, Q_k), B = (b_1, \dots, b_k)$ and $S = (s_{ij})$. Hence, S, being a correlation matrix, is non-singular. So the b's are uniquely estimated by $\hat{B} = QS^{-1}$.

The unconditional minimum is

$$s^2 = L_{min}$$

$$= \sum_i (y_i - \hat{b}_1 x_{1i} - \cdots - \hat{b}_k x_{ki}) y_i$$

(since the other terms vanish by the normal equations)

$$= \sum_i y_i^2 - \hat{b}_1 \sum_i x_{1i} y_i - \cdots - \hat{b}_k \sum_i x_{ki} y_i$$

$$= ns_y^2 - \sum_j \hat{b}_j y_i$$

$$= ns_y^2 - \hat{B} Q^T$$

$$= ns_y^2 - QS^{-1} Q^T$$

$$= ns_y^2 - ns_y^2 R^2$$

$$= ns_y^2 (1 - R^2)$$

where R is the correlation determinant. This has $(n-1) - k$ d.f.. So $\frac{s^2}{\sigma^2} = \frac{ns_y^2(1-R^2)}{\sigma^2}$ is a $\chi^2$ with $(n-k-1)$ d.f..

**To test $H_0$:** $b_1 = \cdots = b_k = 0$

This is a linear hypothesis with k d.f.. The conditional minimum is

$$s_0^2 = \sum y_i^2 = ns_y^2$$

and it has $(n-1)$ d.f.. The test criterion is

$$F(k, n-1-k) = \frac{s_0^2 - s^2/k}{s^2/(n-1-k)} = \frac{R^2/k}{(1-R^2)/(n-1-k)}.$$

**Note:**

As is evident this is an indirect test for the hypothesis that the multiple correlation in the population is zero.

If F is significant, we proceed to test individual hypotheses like $b_i = 0$ and the t-test is the most appropriate one here.

**t-test**

In matrix notation the observational equations are

$$E(y) = BA^T$$

where $B = (b_1, \ldots, b_k)$, $y = (y_1, \ldots, y_n)$ and

$$A = \begin{pmatrix} x_{11} & x_{21} & \cdots & x_{k1} \\ x_{12} & x_{22} & \cdots & x_{k2} \\ \cdots & \cdots & \cdots & \cdots \\ x_{1n} & x_{2n} & \cdots & x_{kn} \end{pmatrix}$$

Since the x's are measured from their means we have $A^T A = S$. The individual parameters are estimable and are given by $\widehat{B} = QS^{-1}$. Also

$$V(\widehat{B}) = \left(A^T A\right)^{-1} \sigma^2 = S^{-1}\sigma^2.$$

An unbiased estimator for $\sigma^2$ is

$$\frac{L_{\min}}{n-r} = \frac{s^2}{(n-1)-k}.$$

Therefore an unbiased estimator of $V(\widehat{B})$ is $S^{-1}\frac{s^2}{(n-1)-k}$. Hence the diagonal elements of $S^{-1}\frac{s^2}{(n-1)-k}$ give unbiased estimates of the variances of $\widehat{b}$'s.

The statistic for testing $H_0: b_i = 0$ is

$$t = \frac{\hat{b}_i}{\sqrt{\text{An unbiased estimate of } V(\hat{b}_i)}}$$

which is a Student t with $(n-1)-1 = (n-2)$ d.f..

Similarly, to test $H_0: b_i = b_{io}$, we use

$$t = \frac{\hat{b}_i - b_{io}}{\sqrt{\text{An unbiased estimate of } V(\hat{b}_i)}}$$

which is a Student t with $(n-2)$ d.f..

**To test $H_0$: $b_i = b_{io}, i = 1,2,\dots,k$.**

This linear hypothesis has k d.f.. The unconditional minimum is

$$s^2 = ns_y^2 - \sum \hat{b}_i Q_i$$

which has $(n-1-k)$ d.f..

The conditional minimum is

$$s_0^2 = \sum_i (y_i - b_{10}x_{1i} - \cdots - b_{k0}x_{ki})y_i$$

$$= ns_y^2 - \sum b_{jo} Q_j$$

which has $(n-1)$ d.f.. So the test statistic in this case is

$$F(k, n-1-k) = \frac{(s_0^2 - s^2)/k}{s^2/(n-1-k)}$$

$$= \frac{\sum_j (\hat{b}_j - b_{jo}) Q_j / k}{(ns_y^2 - \sum \hat{b}_j Q_j)/(n-1-k)}.$$

**To test $\mathbf{H_0}$**: $b_1 = b_2 = \cdots = b_k$ $(= b, \text{say})$

This is a linear hypothesis with $(k-1)$ d.f.. The test statistic can be shown to be equal to

$$F(k, n-1-k) = \frac{(s_0^2 - s^2)/(k-1)}{s^2/(n-1-k)}$$

$$= \frac{\sum_i (\hat{b}_i - \hat{b})\, Q_i/(k-1)}{\left(n s_y^2 - \sum \hat{b}_i Q_i\right)/(n-1-k)}.$$

# Chapter 3

# Analysis of Variance

The Analysis of Variance is a simple arithmetical method of sorting out the components of variation in a given set of results. In order to understand this, it is essential to know what is meant by components of variation.

Whenever there is heterogeneity of variation more than one component is present. Suppose that we have a group of men drawn from a population in which the people are all of the same race and the variable studied is height in inches. A frequency distribution is drawn up and found to be approximately normal. Therefore there is good reason to assume that the variation in this group is approximately homogeneous. The same would apply to a group of women studied in the same manner. However, if the data from the two groups are mixed to form a new group, a second component of variation is brought in, namely the difference between the means of the two groups. This difference would be large enough so that the frequency distribution for the combined groups would probably show two peaks or modes. Carrying the analogy further, a third group might consist of boys from 13 to 15 years of age, and a fourth group of girls of a similar age. When all four group are combined the frequency distribution might appear reasonably normal, but we know that the two components of variation are actually present, one representing variation within the groups and another between the groups. The arithmetical procedure of the analysis of variance enables us to sort out and evaluate the components of variation for such mixed populations.

The complete analysis of variance actually performs a dual role. In the first place we have the sorting-out and estimation of the variance components, and in the second place it provides for tests of significance. The sorting-out process is purely mechanical and can be applied in all cases, but the reliability of the estimates of variance so obtained is dependent to some extent on the manner in which the data are collected.

## 3.1 Fundamental Principles

When we have more than two values we know that the most appropriate measure of the variation among them is the standard deviation or the variance. If $\sigma_{\bar{x}}^2$ is the variance of the means of samples of size n drawn from a population having the variance $\sigma^2$, then we know that the relation between the two variances is expressed as

$$n\sigma_{\bar{x}}^2 = \sigma^2 \tag{1}$$

From this expression a very simple but important deduction can be made. Having drawn a series of samples of size n we can in any event calculate a mean square $V_i$ for any one sample, and each of these can be taken as an estimate of the population variance $\sigma^2$. Furthermore, we can calculate $V(\bar{x})$ from the means of the samples, and from equation (1) above it is clear that $nV(\bar{x})$ can also be taken as an estimate of $\sigma^2$.

In order to clarify the above ideas we shall represent the series of samples by symbols as follows:

| Sample | Variates | Totals | Mean |
|---|---|---|---|
| 1 | $x_{11}$ $x_{12}$ ... $x_{1j}$ ... $x_{1n}$ | ......... | $\bar{x}_1$ |
| 2 | $x_{21}$ $x_{22}$ ... $x_{2j}$ ... $x_{2n}$ | ......... | $\bar{x}_2$ |
| ... | ......... | ......... | ......... |
| i | $x_{i1}$ $x_{i2}$ ... $x_{ij}$ ... $x_{in}$ | ......... | $\bar{x}_i$ |
| ... | ......... | ......... | ......... |
| k | $x_{k1}$ $x_{k2}$ ... $x_{kj}$ ... $x_{kn}$ | ......... | $\bar{x}_k$ |

where there are k samples of n variates each. To give the symbols a concrete meaning it can be supposed that kn rats are divided at random into k samples. Then $x_{ij}$ is the weight of the j-th rat in the i-th sample.

For each sample we calculate

$$V_i = \frac{1}{n-1}\sum_{j=1}^{n} x_{ij}^2 \tag{2}$$

where $x_{ij} = x_{ij} - \bar{x}_i$. Since each of these is an unbiased estimate of $\sigma^2$, the series can be averaged to provide a single estimate. Thus

$$V = \frac{1}{k}\sum_{i=1}^{k} V_i = \frac{1}{k(n-1)}\sum_{i=1}^{k}\sum_{j=1}^{n} x_{ij}^2 \quad (3)$$

which amounts to summing all the sums of squares of deviations from sample means and dividing by the total number of degrees of freedom available within samples.

To find $nV(\bar{x})$ we take

$$nV(\bar{x}) = n\frac{1}{k-1}\sum_{i=1}^{k}(\bar{x}_i - \bar{x})^2 \quad (4)$$

where $\bar{x}$ is the average of all sample means.

If the variates in all samples are drawn at random, both V and $nV(\bar{x})$ are unbiased estimates of $\sigma^2$, but they are of course not equally reliable since V is based on $k(n-1)$ d.f. and $nV(\bar{x})$ on only $(k-1)$ d.f.. The random arrangement ensures that one rat has an equal chance of being included in any sample. It should be clear, therefore, that the only factor affecting the values of both V and $nV(\bar{x})$ is the variability of the population from which the samples are drawn.

Our main interest in the proposition outlined above lies in the situation that is created when the population is not homogeneous, as in the first example where the variance due to means is affected by fundamental differences between the groups. With groups of rat, each group may have been given a different ration. The conditions causing variation within the groups will be due to the original variation in the population, but the means of the samples will vary additionally owing to the differences in the rations. With symbols the situation can be represented as follows:

| Sample | Variates | Sample Mean | Mean Square |
|---|---|---|---|
| 1 | $x_{11} + y_1$ $x_{12} + y_1$ ... $x_{1n} + y_1$ | $\bar{x}_1 + y_1$ | $V_1$ |
| 2 | $x_{21} + y_2$ $x_{22} + y_2$ ... $x_{2n} + y_2$ | $\bar{x}_2 + y_2$ | $V_2$ |
| ... | ……… | ……… | ……… |
| k | $x_{k1} + y_k$ $x_{k2} + y_k$ ... $x_{kn} + y_k$ | $\bar{x}_k + y_k$ | $V_k$ |

Note that in sample 1 each variate value consists of a part x, as in the case where all variates are drawn at random, plus a portion y, which is constant for each variate. Thus the mean of sample 1 is $\frac{\sum x}{n} + \frac{ny_1}{n} = \bar{x}_1 + y_1$. Since y is constant for any one sample, it does not affect the value of $V_i$. Algebraically, we have, for sample 1,

$$V_1 = \frac{1}{n-1}\sum_1^n [(x_1 + y_1) - (\bar{x}_1 + y_1)]^2 = \frac{1}{n-1}\sum_1^n (x_1 - \bar{x}_1)^2$$

$$= \frac{1}{n-1}\sum_1^n x_1^2. \quad (5)$$

Therefore V based on all samples will be the same as if the y factor did not exist.

Examination of the new mean square for sample means shows that it is definitely affected by differences among the values of y. We shall represent it by $V'(\bar{x})$.

$$V'(\bar{x}) = \frac{1}{k-1}\sum_1^k [(\bar{x}_i + y_i) - (\bar{x} + \bar{y})]^2$$

$$= \frac{1}{k-1}\sum_1^k [(\bar{x}_i - \bar{x}) + (y_i - \bar{y})]^2$$

$$= \frac{1}{k-1}\sum_1^k (\bar{x}_i - \bar{x})^2 + \frac{1}{k-1}\sum_1^k (y_i - \bar{y})^2$$

$$+ \frac{2}{k-1}\sum_1^k (\bar{x}_i - \bar{x})(y_i - \bar{y})$$

$$= V(\bar{x}) + V(y) + \frac{2}{k-1}\sum_1^k (\bar{x}_i - \bar{x})(y_i - \bar{y}), \quad (6)$$

thus $V'(\bar{x})$ consists row of $V(\bar{x}) + V(y)$ plus the additional quantity which is a sum of products of deviations. Now the part of the differences between the sample means and the general mean represented by $\bar{x}_i - \bar{x}$

arises purely from random sampling and would bear no relation to corresponding values of $y_i - \bar{y}$. Therefore, we can assume that for a large series of samples this sum of products will be zero. We are left with

$$V'(\bar{x}) = V(\bar{x}) + V(y) \text{ or}$$

$$nV'(\bar{x}) = nV(\bar{x}) + nV(y) \qquad (7)$$

and conclude that for such a series of samples $nV'(\bar{x})$ will tend to be greater than $nV(\bar{x})$ by an amount equal to $nV(y)$.

In an actual experiment we do not know the value of $V(\bar{x})$ because we do not have a set of figures showing the random variation only. All we have are the final figures containing both x and y. We know, however, that V is a good estimate of $nV(\bar{x})$, therefore from (7),

$$nV(y) = nV'(\bar{x}) - V, \text{ see equation (3)} \qquad (8)$$

gives a good estimate of the portion of the total variance due to means that can be attributed to the y effects arising from the treatments applied to the samples.

This procedure is known as estimating the variance component due to treatments and represents one of the important functions of the analysis of variance.

**Example 3.1**

A numerical example will illustrate the procedure clearly. The figures below have been placed at random into 5 groups of 5.

| Group | Variates | Mean |
|---|---|---|
| 1 | 29 45 14 25 11 | 24.8 |
| 2 | 44 38 48 31 31 | 38.4 |
| 3 | 16 29 29 18 46 | 27.6 |
| 4 | 37 11 31 28 44 | 30.2 |
| 5 | 50 12 19 20 28 | 25.8 |

Making calculations, $V = 150.34$ and then $nV(\bar{x}) = 148.74$. The agreement here is better than would ordinarily be expected. In a large series of such samples we would, however, expect to get very close agreement between V and $nV(\bar{x})$.

Now, let us take $y_1 = 0$, $y_2 = 0$, $y_3 = 0$, $y_4 = 0$, and $y_5 = -2$. Then the figures are

| Group | Variates | Mean |
|---|---|---|
| 1 | 29 45 14 25 11 | 24.8 |
| 2 | 46 40 50 33 33 | 40.4 |
| 3 | 16 29 29 18 46 | 27.6 |
| 4 | 37 11 31 28 44 | 30.2 |
| 5 | 48 10 17 18 26 | 23.8 |

We get $V = 150.34$ as before and $nV(\bar{x}) = 221.74$. The latter is considerably inflated over the previous figure of 148.74, showing the effect of y.

Therefore a good estimate of the portion of the total variance due to means that can be attributed to the y effects is

$$221.74 - 150.34 = 71.40.$$

The same general principles hold when the data are classified in two ways, as in a simple replicated experiment with a series of treatments. The data may be arranged in a table as follows:

Treatment

Replicate

| R \ T | 1 | 2 | ... | j | ... | n | Means |
|---|---|---|---|---|---|---|---|
| 1 | $x_{11}$ | $x_{12}$ | ... | ... | ... | $x_{1n}$ | $\bar{R}_1$ |
| 2 | $x_{21}$ | $x_{22}$ | ... | ... | ... | $x_{2n}$ | $\bar{R}_2$ |
| $\vdots$ | $\vdots$ | $\vdots$ | $\vdots$ | $\vdots$ | $\vdots$ | $\vdots$ | $\vdots$ |
| i | ... | ... | ... | $x_{ij}$ | ... | ... | $\vdots$ |
| $\vdots$ | $\vdots$ | $\vdots$ | $\vdots$ | $\vdots$ | $\vdots$ | $\vdots$ | $\vdots$ |
| r | $x_{r1}$ | $x_{r2}$ | ... | ... | ... | $x_{rn}$ | $\bar{R}_r$ |
| Means | $\bar{T}_1$ | $\bar{T}_2$ | ... | ... | ... | $\bar{T}_n$ | M |

Here $x_{ij}$ represents the yield of treatment j in replicate 1. This table may be regarded as made up of r samples of n each, or n samples of r each. Thus, if the variates were all drawn at random from a population with

variance $\sigma^2$ and simply arranged as above in rows and columns, we could make and estimate of this variance in three ways:
(i) from the means of the rows,
(ii) from the means of the columns, and
(iii) from the residual mean square not affected by rows and columns.

In order to clarify this point it is convenient to represent each yield by $x + y + z$ instead of by x alone. The y is the portion representing the replicates and z the portion representing the treatment.

The same reasoning as was applied to the example of k samples of n each will then show that an estimate of the population variance from the means of the replicates gives

$$nV'(\overline{R}) = nV(\overline{R}) + nV(y) \tag{9}$$

and from the means of the treatments

$$nV'(\overline{T}) = nV(\overline{T}) + nV(z) \tag{10}$$

In (9) $nV(y)$ is that portion of the mean square for replicates that is due to real variation among the replicates and not to random variation from plot to plot. Also in (10) $nV(z)$ is the component of the mean square due only to treatments.

When each variate is represented by $x + y + z$, where $y_1$ will be the contribution due to replicated 1 and $z_1$ the contribution due to treatment 1, the first variate can be represented by $x_{11} + y_1 + z_1$. Then, in order to obtain an estimate of random variation only it should be noted that

$$(x_{11} + y_1 + z_1 - M) - [(\bar{x}_{R1} + y_1 + \bar{z}) - M] \\ -[(\bar{x}_{T1} + y_1 + \bar{z})M] \tag{11}$$

where M is the general mean, $\bar{x}_{R1}$ the mean of the replicate containing $x_{11}$, and $\bar{x}_{T1}$ the mean of the treatment containing $x_{11}$, can be simplified to

$$(x_{11} - \bar{x}) - (\bar{x}_{R1} - \bar{x}) - (\bar{x}_{T1} - \bar{x}) \tag{12}$$

since $M = \bar{x} + \bar{y} + \bar{z}$. This shows that, for and individual deviation from the general mean, if the deviations due to replicate and treatment means are subtracted, the remaining portion represents random variation only. It shows how a mean square can be calculated that will be an estimate of $\sigma^2$ in that all effects due to replicates and treatments are removed.

## 3.2 Partitioning of Sums of Squares and Degrees of Freedom

Suppose that we have a table of results as follows:

| Replicate: R \ Treatment: T | 1 | 2 | ⋯ | j | ⋯ | n | Replicate Totals |
|---|---|---|---|---|---|---|---|
| 1 | $x_{11}$ | $x_{12}$ | ⋯ | ⋯ | ⋯ | $x_{1n}$ | $R_1$ |
| 2 | $x_{21}$ | $x_{22}$ | ⋯ | ⋯ | ⋯ | $x_{2n}$ | $R_2$ |
| ⋮ | ⋮ | ⋮ | ⋮ | ⋮ | ⋮ | ⋮ | ⋮ |
| i | ⋯ | ⋯ | ⋯ | $x_{ij}$ | ⋯ | ⋯ | ⋮ |
| ⋮ | ⋮ | ⋮ | ⋮ | ⋮ | ⋮ | ⋮ | ⋮ |
| r | $x_{r1}$ | $x_{r2}$ | ⋯ | ⋯ | ⋯ | $x_{rn}$ | $R_r$ |
| Treatment Totals | $C_1$ | $C_2$ | ⋯ | ⋯ | ⋯ | $C_n$ | G = Grand Total |

For any one observation, say $x_{11}$, we can write

$$x_{11} - \bar{x} = (x_{11} - \bar{R}_1) + (\bar{R}_1 - \bar{x})$$

where $\bar{x}$ is the general mean and $\bar{R}_1$ is the mean of replicate 1. Then

$$(x_{11} - \bar{x})^2 = (x_{11} - \bar{R}_1)^2 + (\bar{R}_1 - \bar{x})^2 + 2(x_{11} - \bar{R}_1)(\bar{R}_1 - \bar{x})$$

and, summing over replicate 1,

$$\sum_{1}^{n}(x - \bar{x})^2 = \sum_{1}^{n}(x - \bar{R}_1)^2 + n(\bar{R}_1 - \bar{x})^2 + 2(\bar{R}_1 - \bar{x})\sum_{1}^{n}(x_1 - \bar{R}_1).$$

The last term is zero. Thus we have

$$\sum_{1}^{n}(x - \bar{x})^2 = \sum_{1}^{n}(x - \bar{R}_1)^2 + n(\bar{R}_1 - \bar{x})^2.$$

Now, if we repeat this for each replicate and sum over the whole experiment, we get

$$\sum_{1}^{rn}(x-\bar{x})^2 = \sum_{1}^{r}\sum_{1}^{n}(x-\bar{R}_r)^2$$

$$+n\sum_{1}^{r}(\bar{R}_r-\bar{x})^2. \qquad (1)$$

This is a very important equation in that it shows how the total sum of squares can be partitioned into one part representing deviations from the mean within replicates and another part representing deviations of the replicate means from the general mean. Simply by extending the algebra we can show that, if a deviation of $x_{11}$ from the general mean is written as

$$x_{11}-\bar{x} = (\bar{R}_1-\bar{x}) + (\bar{C}_1-\bar{x}) + [(x_{11}-\bar{x})-(\bar{R}_1-\bar{x})-(\bar{C}_1-\bar{x})]$$
$$= (\bar{R}_1-\bar{x}) + (\bar{C}_1-\bar{x}) + (x-\bar{R}_1-\bar{C}_1+\bar{x})$$

then the total sum of squares can be partitioned into

$$\sum_{1}^{rn}(x-\bar{x})^2 = n\sum_{1}^{r}(\bar{R}-\bar{x})^2 + r\sum_{1}^{n}(\bar{C}-\bar{x})^2$$

$$+\sum_{1}^{rn}(x-\bar{R}_i-\bar{C}_j+\bar{x})^2 \qquad (2)$$

(Total = Replicates + Treatments + Random Variation)

where i runs from 1 to r and j from 1 to n.

The degrees of freedom can be partitioned in a similar manner. Adjustment for the mean takes up one d. f.; therefore a total of $rn-1$ are available for partitioning. Taking $(r-1)$ d. f. for replicates and $(n-1)$ for treatments leaves

$$(rn - 1) - (r - 1) - (n - 1) = (r - 1)(n - 1)$$

for random variations. Therefore the equation for d. f. corresponding to (2) is

$$rn - 1 = (r - 1) - (n - 1) + (r - 1)(n - 1).$$

$$(\text{Total} = \text{Replicates} + \text{Treatments} + \text{Random Variations})$$

The complete analysis of variance is most conveniently set up in tabular form as shown below, where we now substitute the term error for random variation.

| Source | Sum of Squares | d. f. | Mean Square |
|---|---|---|---|
| Replicates | $n\sum(\bar{R} - \bar{x})^2$ | $r - 1$ | $nV'(\bar{R})$ |
| Treatments | $n\sum(\bar{C} - \bar{x})^2$ | $n - 1$ | $nV'(\bar{C})$ |
| Error | $\sum(x - \bar{R} - \bar{C} + \bar{x})^2$ | $(r - 1)(n - 1)$ | $V$ |
| Total | $\sum(x - \bar{x})^2$ | $rn - 1$ | |

This form is convenient for tabulation and for tests of significance. An F value for replicates or for treatments can be calculated from

$$F_r = \frac{nV'(\bar{R})}{V}, \quad F_C = \frac{nV'(\bar{C})}{V}.$$

For $F_r$ the degrees of freedom are $(r - 1)$ and $(r - 1)(n - 1)$, and for treatments they are $(n - 1)$ and $(r - 1)(n - 1)$.

The calculations are best carried out as in the formulas given below:

Total $$\sum_{1}^{rn}(x-\bar{x})^2 = \sum_{1}^{rn} x^2 - \frac{G^2}{rn} \quad (3)$$

Replicates $$n\sum_{1}^{r}(\bar{R}-\bar{x})^2 = \frac{1}{n}\sum_{1}^{r} R^2 - \frac{G^2}{rn} \quad (4)$$

Treatments $$r\sum_{1}^{n}(\bar{C}-\bar{x})^2 = \frac{1}{r}\sum_{1}^{n} C^2 - \frac{G^2}{rn} \quad (5)$$

Error = Total – Replicates – Treatments.

The entire procedure, including tests of significance, is given in the example below.

**Example 3.2 Two-fold classification of variates**

In a swine feeding experiment Dunlop obtained the results given in the table below. The three rations, A, B, and C, differed in the substances providing the vitamins. The animals were in 4 groups of 3 each, the grouping being on the basis of litter and initial weights. For our purpose we shall assume that the grouping is merely a matter of replication.

Table: Gains in weight of swine fed on rations A, B, C.

| Ration \ Group | I | II | III | IV | Total |
|---|---|---|---|---|---|
| A | 7.0 | 16.0 | 10.5 | 13.5 | 47.0 |
| B | 14.0 | 15.5 | 15.0 | 21.0 | 65.5 |
| C | 8.5 | 16.5 | 9.5 | 13.5 | 48.0 |
| Total | 29.5 | 48.0 | 35.0 | 48.0 | 160.5 |

The form of the analysis is

| S.S. | d.f. |
|---|---|
| Rations | 2 |
| Groups | 3 |
| Error | 6 |
| Total | 11 |

Calculating the sums of squares, we have

$$\text{Total} = 2316.75 - \frac{(160.5)^2}{12} = 170.06$$

$$\text{Rations} = \frac{(47.0)^2 + (65.5)^2 + (48.0)^2}{4} - 2146.69 = 54.12$$

$$\text{Groups} = \frac{(29.5)^2 + \cdots + (48.0)^2}{3} - 2146.69 = 87.73$$

$$\text{Error} = 170.06 - 54.12 - 87.73 = 28.21.$$

This gives an analysis of variance as follows:

| Source | S. S. | d. f. | M. S. | F. | 5% point |
|---|---|---|---|---|---|
| Rations | 54.12 | 2 | 27.06 | 5.75 | 5.14 |
| Groups | 87.73 | 3 | 29.24 | 6.22 | 4.76 |
| Error | 28.21 | 6 | 4.702 | | |
| Total | 170.06 | 11 | | | |

The mean square for rations is just significant. The meaning of the significance of the mean square for groups depends on the manner in which the classification into groups has been made. We have assumed here that the groups are merely replications, in which case the error mean square is a result of variations within groups not due to the rations. It is, therefore, valid to consider this mean square as an error with which the others can be compared. The group mean square, since it results from the plan of the experiment, is an expression of error control. If the arrangement had been other than in groups, we would have had a simple classification into within and between rations. Then mean square for within rations would have been much larger than it is according to the present arrangement, and consequently the experiment would have been less precise.

## 3.3 Experimental Error

In previous discussions the variates have been presented by symbols such as $x_{11} + y_1 + z_1$ where $x_{11}$ is the portion due entirely to random variability, and our problem in the analysis of variance was to isolate the mean square resulting from this random variability from other mean squares resulting from other mean squares effects of treatment, replicates, etc. This mean square due to random variability is commonly referred to as the mean square for experimental error. In other words, in an experiment it represents that portion of the total variability that is beyond control. If the previous discussion has been followed carefully, it will be obvious that the experimental error is the logical measuring stick for making tests of significance of other mean squares, because, if the mean square tested is not inflated by any real effect in the experiment, it must tend to be equal to the experimental error.

## 3.4 Assumptions Underlying the Analysis of Variance

First developed by Fisher as a method of analyzing the variation to which experimental and observational material is subject to, by differentiating the variation according to causes or groups of causes – assessing the various components of variation – analysis of variance is still the only efficient method to achieve this purpose. Besides serving as a simple arithmetical procedure for arranging and presenting the experimental results in a simple compact table, illustrating the structure of the experiment, analysis of variance serves double propose of detection and estimation of constant relations among the means of subsets and the detection and estimation of the components of random variation associated with a composite population.

Before using the analysis of variance to summarize the results of an experiment, it is advisable to check the reasonableness of the assumptions which are set up. When the formulas and procedures of analysis of variance are used merely to summarize properties of the data in hand, no assumptions are required to validate them. On the other hand when analysis of variance is used as a method of statistical inference – for inferring properties of the population from which the data at hand were drawn – then certain assumptions about the population and the sampling procedure by means of which the data were obtained must be fulfilled if the inferences are to be valid.

**(i) Random Variables**

The observations are values of random variables that are distributed about a fixed, true but unknown mean.

**(ii) Additivity**

In order to connect the analysis of variance with the theory of linear estimation, we must assume that the various fixed effects and the error are additively related to the true mean. If the data is such that the effects are really not additive, then the sum of squares attributable to such effects by the analysis of variance does not represent true effects. Further any contrast between effects of two items of classification will be so only when the other factors are as in the experiment concerned. Hence additivity does not prevail, we say that there are interactions between the different classifications. If this interaction is also additive, then the analysis of variance may be set up to measure this too. Suitable transformations may be effected to make non – additive models additive.

**(iii) Non – Correlated Error**

The error of any observation must be uncorrelated. (not necessarily independent) with that for any other observation. But usually this assumption is violated. In this case proper randomization will help to overcome this difficulty, not because it removes any correlation present, but because it provides a mechanism by which this expected correlation between two treatments tends to cancel as the number of experimental units per treatment is increased.

**(iv) Homogeneity of Variance for Simplicity of Analysis of Variance**

In order to have a simple analysis of variance it is desirable that the errors be the same from one experimental unit to another, regardless of the treatment used. If the errors are not equal it is necessary to know the relative sizes of the variances which is generally not available. Sometimes the data can be split into parts, each part with homogeneous errors but with unequal errors from one part to another.(as in a split – plot design.) In order for variation to be strictly homogeneous, it must be purely random, caused by a multiplicity of independent factors, incapable of resolution into more elemental form and indistinguishable from one another. Bartlett discusses various transformations of the original data to stabilize the variance when there is a fixed relationship between the mean and the variance, like the logarithmic and square root transformations. However, if several differences in effects are small, there is seldom any need for a transformation.

### (v) Normality for Valid Tests of Significance

If it is desired to make exact tests of significance or set up confidence limits, the errors must be normally distributed. If this condition is satisfied then the previous condition becomes equivalent to independence. However, the analysis of variance tests may be considered to be approximate tests (rather than exact tests) without serious error even when the assumption of normality is not met.

Thus the necessary and sufficient conditions for the strict validity of analysis of variance procedures are the additive model with random errors which are normally and independently distributed with zero mean and preferably a common variance.

### Analysis of Variance Table

It is a neat summary form of the results of an experiment showing the various components of variation such that they can be subjected to statistical tests of hypotheses and the structure will be

| Variation Due to | Sum of Squares (a) | Degrees of freedom (b) | Mean Sum of Squares $\frac{(a)}{(b)}$ | F-ratio |
|---|---|---|---|---|
| ... | ... | ... | ... | ... |
| ... | ... | ... | ... | ... |
| Residual (error) | (By subtraction) | (By subtraction) | | |
| Total | | | | |

## 3.5 Analysis of Variance in the Case of Regression

### (i) Simple Regression

The model is $E(y) = a + bx$ and the least square estimates of a and b are given by

$$\hat{a} = \bar{y} - \hat{b}\bar{x} \quad \text{and} \quad \hat{b} = \frac{\sum(x-\bar{x})(y-\bar{y})}{\sum(x-\bar{x})^2}.$$

Suppose we want to test the hypothesis that $b = 0$, i. e., there is no regression. When there is no regression, the total error sum of squares is $\sum(y-\bar{y})^2$. When there is regression, the minimum error sum of squares is

$$\sum(y-\bar{y})^2 - \hat{b}\sum(x-\bar{x})(y-\bar{y}).$$

Therefore the part due to regression is

$$\hat{b}\sum(x-\bar{x})(y-\bar{y}).$$

In this case the Analysis of Variance table will be as follows:

| Variation Due to | S.S. (i) | d.f. (ii) | M.S.S (i)/(ii) | F-ratio |
|---|---|---|---|---|
| Regression | $\hat{b}\sum(x-\bar{x})(y-\bar{y})$ | 1 | $V_R$ | $\frac{V_R}{V_E} = F(1, n-2)$ |
| Residual | (Sub) | $n-2$ | $V_E$ | |
| Total | $\sum(y-\bar{y})^2$ | $n-1$ | | |

### (ii) Multiple Regression

The model is

$$E(y) = b_1x_1 + \cdots + b_kx_k.$$

Using our earlier notations, estimates of $b_1, \ldots, b_k$ are given by

$$\hat{B} = (\hat{b}_1, \ldots, \hat{b}_k) = Q^TD^{-1}.$$

The unconditional minimum is

$$S^2 = \sum y^2 - \sum \hat{b}_i Q_i$$

Under $H_0: b_1 = \cdots = b_k = 0$, the conditional minimum is

$$S_0^2 = \sum y^2.$$

Therefore the variation due to regression is $\sum \hat{b}_i Q_i$. We can set up the analysis of variance table as follows:

| Variation Due to | S.S. (i) | d.f.(ii) | M.S.S. (i)/(ii) | F-ratio |
|---|---|---|---|---|
| Regression | $\sum \hat{b}_i Q_i$ | $k$ | $V_R$ | $\frac{V_R}{V_E} = F(k, n-k)$ |
| Residual | (Sub) | $n-k$ | $V_E$ | |
| Total | $\sum y^2$ | $n$ | | |

**(iii) Partial Regression**

A frequent problem in regression is the following. We wish to test whether certain regression coefficients are zero without making any assumptions about the remaining coefficients. Suppose we rename our regression coefficients as $b_1, \dots, b_q, b_{q+1}, \dots, b_k$ and we wish to test $H_0: b_{q+1} \dots = b_k = 0$. The procedure is to split the total variation into two groups: (i) the variation due to the first q variables, and (ii) the variation due to all the k variables.

The least square estimates of the k parameters are given by $\hat{B} = Q^T D^{-1}$ where $\hat{B} = (\hat{b}_1, \dots, \hat{b}_k)$ and Q and D have meanings explained before. Then the unconditional minimum will be

$$S^2 = \sum y^2 - \sum_1^k \hat{b}_i Q_i$$

which has $(n-k)$ d.f..

Now let us estimate $b_1, \dots, b_q$ under $H_0$. The error sum of squares is

$$L = \sum (y - b_1 x_1 - \cdots - b_q x_q)^2.$$

The minimizing conditions are $\frac{\partial L}{\partial b_i} = 0; i = 1, \ldots, q$. These give the normal equations as

$$Q_1 = b_1 S_{11} + \cdots + b_q S_{q1}$$

$$\ldots \ldots \ldots$$

$$Q_q = b_1 S_{1q} + \cdots + b_q S_{qq}.$$

If $b'_1, \ldots, b'_q$ are the solutions, the conditional minimum will be

$$S_0^2 = \sum y^2 - \sum_1^k b'_i Q_i$$

with $(n - q)$ d.f..

Therefore variation due to regression on $x_{q+1}, \ldots, x_k$ is

$$S_0^2 - S^2 = \sum_1^k \hat{b}_i Q_i - \sum_1^k b'_i Q_i$$

and this has $(k - 1)$ d.f..

The analysis of variance table is shown below.

| Variation Due to | S.S. | d.f. | M.S.S | F-ratio |
|---|---|---|---|---|
| (1) Regression on $x_1, \ldots, x_q$ | $\sum_1^k b'_i Q_i$ | q | $V_q$ | |
| (2) Regression on $x_{q+1}, \ldots, x_q$ after removing the effect due to regression on $x_1, \ldots, x_q$ | (3) − (2) | k − q | V | $\frac{V}{V_E} = F(k-q, n-k)$ |
| (3) Regression on $x_1, \ldots, x_k$ | $\sum_1^k \hat{b}_i Q_i$ | k | $V_k$ | |
| (4) Residual | (5) − (4) | n − k | $V_E$ | |
| (5) Total | $\sum y^2$ | n | | |

### 3.6 One–Way Classification

Let there be k independent samples of sizes $n_1, \dots, n_k$ from k populations with unknown means $\alpha_1, \dots, \alpha_k$ and with a common unknown variance $\sigma^2$. Assuming the additive model

$$y_{ij} = \alpha_i + \varepsilon_{ij}; \quad j = 1, \dots, n_i; i = 1,2, \dots, k$$

we have $E(\varepsilon_{ij}) = 0$ and $V(\varepsilon_{ij}) = \sigma^2$. The total error sum of squares is

$$L = \sum_{i,j} \varepsilon_{ij}^2 = \sum_{i,j} (y_{ij} - \alpha_i)^2.$$

The minimizing conditions are $\frac{\partial L}{\partial \alpha_i} = 0; i = 1, \dots, k,$

$$\text{i.e.,} \sum_j (y_{ij} - \alpha_i) = 0 \text{ or } y_{i\cdot} - n_i\alpha_i = 0.$$

Therefore the least square estimate of $\alpha_i$ is

$$\hat{\alpha}_i = \frac{y_{i\cdot}}{n_i} = \bar{y}_{i.} \,.$$

So the conditional minimum is

$$S^2 = \sum_{i,j} (y_{ij} - \hat{\alpha}_i)^2$$

$$= \sum_{i,j} (y_{ij} - \hat{\alpha}_i) y_{ij} \text{, by the normal euqations}$$

$$= \sum_{i,j} (y_{ij} - \bar{y}_{i\cdot}) y_{ij}$$

$$= \sum_{i,j} y_{ij}^2 - \sum_i \bar{y}_{i\cdot} y_{i\cdot}$$

$$= \sum_{i,j} y_{ij}^2 - \sum_i n_i \bar{y}_{i.}^2 \,.$$

Suppose we want to test the hypothesis $H_0: \alpha_1 = \cdots = \alpha_k$, a linear hypothesis with $(k - 1)$ d.f.. Let under $H_0$, $\alpha_1 = \cdots = \alpha_k = \alpha$. Then we have to minimize

$$L_0 = \sum_{i,j} (y_{ij} - \alpha)^2.$$

The least square normal equation is

$$\sum_{i,j}(y_{ij} - \alpha) = 0,$$
$$\text{i.e.}, y_{..} - n.\alpha = 0$$

giving the least square estimate of $\alpha$ to be

$$\hat{\alpha} = \frac{y_{..}}{n.} = \bar{y}_{..}\,.$$

Then the conditional minimum then is

$$S_0^2 = \sum_{i,j}(y_{ij} - \hat{\alpha})y_{ij} = \sum_{i,j} y_{ij}^2 - n.\bar{y}_{..}^2,$$

which has $n. - k + k - 1 = (n. - 1)$ d.f..

Note that $S^2$ has $(n. - k)$ d.f. and so $S_0^2 - S^2$ which is the sum of squares due to deviation from the hypothesis has $(k - 1)$ d.f.. Under normality assumption the test criterion is

$$F(k - 1, n. - k) = \frac{(S_0^2 - S^2)/(k - 1)}{S^2/(n. - k)}$$
$$= \frac{(\sum_i n_i \bar{y}_{i.}^2 - n.\bar{y}_{..}^2)/(k - 1)}{(\sum_{i,j} \bar{y}_{ij}^2 - \sum_i n_i \bar{y}_{i.}^2)/(n. - k)}.$$

If this F is significant we reject $H_0$ and proceed to test individual hypothesis such as $H_0': \alpha_i = \alpha_j$. Under $H_0'$, the error sum of squares is

$$L_0' = \sum_r (y_{1r} - \alpha_1)^2 + \cdots + \sum_r (y_{ir} - \alpha_i)^2 + \cdots + \sum_r (y_{jr} - \alpha_i)^2$$
$$+ \cdots + \sum_r (y_{kr} - \alpha_k)^2\,.$$

The least square estimates of $\alpha_1, \ldots, \alpha_{i-1}, \alpha_{i+1}, \ldots, \alpha_{j-1}, \alpha_{j+1}, \ldots, \alpha_k$ will be the same as before. The normal equation giving $\alpha_i$ is $\frac{\partial L}{\partial \alpha_i} = 0$

$$\text{i.e.}, \sum_r (y_{ir} - \alpha_i) + \sum_r (y_{jr} - \alpha_i) = 0$$

from which we get $\hat{\alpha}_i = \frac{y_{i.} + y_{j.}}{n_i + n_j}$.

Using these estimates we can find the conditional minimum ${S'_0}^2$ and apply the F-test. But a more apt test in the case of such individual hypotheses will be the t-test. We have $\widehat{\alpha}_i = \bar{y}_{i\cdot}$ and $\widehat{\alpha}_j = \bar{y}_{j\cdot}$ and under $H'_0$,

$$t = \frac{\widehat{\alpha}_i - \widehat{\alpha}_j}{\sqrt{\text{unbiased estimate of } V(\widehat{\alpha}_i - \widehat{\alpha}_j)}}$$

is a Student's t with $(n. - k)$ d. f.. Now,

$$\begin{aligned} V(\widehat{\alpha}_i - \widehat{\alpha}_j) &= V(\bar{y}_{i\cdot} - \bar{y}_{j\cdot}) \\ &= V(\bar{y}_{i\cdot}) + V(\bar{y}_{j\cdot}) \\ &= \frac{\sigma^2}{n_i} + \frac{\sigma^2}{n_j} \\ &= (\frac{1}{n_i} + \frac{1}{n_j})\sigma^2 \end{aligned}$$

and an unbiased estimate of $\sigma^2$ is $\frac{S^2}{n.-k}$ where $S^2 = L_{min}$. Thus

$$t = \frac{\bar{y}_{i\cdot} - \bar{y}_{j\cdot}}{\sqrt{(\frac{1}{n_i} + \frac{1}{n_j})\frac{S^2}{n. - k}}}.$$

The construction of the analysis of variance table will be as given below.

The total sum of squares is

$$\begin{aligned} \sum_{i,j}(y_{ij} - \bar{y}_{..})^2 &= \sum_{i,j}(y_{ij} - \bar{y}_{i.} + \bar{y}_{i.} - \bar{y}_{..})^2 \\ &= \sum_{i,j}(y_{ij} - \bar{y}_{i.})^2 + \sum_{i,j}(\bar{y}_{i.} - \bar{y}_{..})^2 \\ &\quad + 2\sum_{i,j}(y_{ij} - \bar{y}_{i.})(\bar{y}_{i.} - \bar{y}_{..}) \end{aligned}$$

$$= \sum_{i,j} (y_{ij} - \bar{y}_{i.})^2$$

$$+ \sum_{i,j} (\bar{y}_{i.} - \bar{y}_{..})^2 \text{, since the last term is zero}$$

$$= \sum_{i,j} (y_{ij} - \bar{y}_{i.})^2 + \sum_{i} n_i (\bar{y}_{i.} - \bar{y}_{..})^2.$$

Now, $\sum_j (y_{ij} - \bar{y}_{i.})^2$ gives the variation in the i-th group. So $\sum_{i,j} (y_{ij} - \bar{y}_{i.})^2$ gives the sum of the within group variation and it has $(n_. - k)$ d.f.. Similarly, $\sum_i n_i (\bar{y}_{i.} - \bar{y}_{..})^2$ gives the sum of the between group variation and it has $(k-1)$ d.f.. Thus the total S.S. with d.f. $(n_. - 1)$ has been split into the S.S. within groups with d.f. $(n_. - k)$ and S.S. between groups with d.f. $(k-1)$. (All these three quantities when divided by the respective d.f. will give an unbiased estimate of $\sigma^2$.) Again,

$$\sum_{i,j} (y_{ij} - \bar{y}_{i.})^2 = \sum_{i,j} (y_{ij}^2 + \bar{y}_{i.}^2 - 2 y_{ij} \bar{y}_{i.})$$

$$= \sum_{i,j} y_{ij}^2 - \sum_{i} \frac{y_{i.}^2}{n_i}, \quad \text{(on simplification)}$$

Also,

$$\sum_{i} n_i (\bar{y}_{i.} - \bar{y}_{..})^2 = \sum_{i} n_i (\bar{y}_{i.}^2 + \bar{y}_{..}^2 - 2 \bar{y}_{i.} \bar{y}_{..})$$

$$= \sum_{i} \frac{y_{i.}^2}{n_i} - \frac{y_{..}^2}{n_i}$$

and

$$\sum_{i,j} (\bar{y}_{i.} - \bar{y}_{..})^2 = \sum_{i,j} y_{ij}^2 - \frac{y_{..}^2}{n_i}.$$

Noting that the within group variation is the same as the residual, we can set up the analysis of variance table as follows:

| Variation due to | S.S. | d.f. | M.S.S. | F |
|---|---|---|---|---|
| Between group | $\sum_i \frac{y_{i.}^2}{n_i} - \frac{y_{..}^2}{n_.}$ | $k-1$ | $V_B$ | $\frac{V_B}{V_E} = F(k-1, n_. - k)$ |
| Within group (Residual) | (By sub) | (By sub) | $V_E$ | |
| Total | $\sum_{i,j} y_{ij}^2 - \frac{y_{..}^2}{n_.}$ | $n_. - 1$ | | |

If F is significant, the variation between groups is significant and so we cannot conclude that the group effects are the same.

Computational procedure can be summarized below:

| Sample | Observations | Sum | S.S. | M.S.S. |
|---|---|---|---|---|
| 1 | $y_{11}\ y_{12} \cdots y_{1n_1}$ | $y_{1.}$ | $y_{1.}^2$ | $y_{1.}^2/n_1$ |
| 2 | $y_{21}\ y_{22} \cdots y_{2n_2}$ | $y_{2.}$ | $y_{2.}^2$ | $y_{2.}^2/n_2$ |
| … | … … | … … | … … | … … |
| k | $y_{k1}\ y_{k2} \cdots y_{kn_k}$ | $y_{k.}$ | $y_{k.}^2$ | $y_{k.}^2/n_k$ |
| Sum | | $y_{..}$ | | $\sum_i \frac{y_{i.}^2}{n_i}$ |

Also $\sum_{i,j} y_{ij}^2$ and the "Correction factor" $\frac{y_{..}^2}{n_.}$ are easily obtained.

**Example 3.3**

Consider an agricultural experiment for testing the effects of different varieties of seeds on the yield. Let $V_1, \ldots, V_k$ be k varieties of seeds. We conduct $n_. = n_1 + n_2 + \cdots + n_k$ experiments, $n_1$ with $V_1$, $n_2$ with $V_2$, etc. and $n_k$ with $V_k$ on $n_.$ unit plots. By properly designing the experiment we can control all other effects except that due to variety. Then the yield in each variety will be the yield due to the effect of that variety together with a random error whose cause is not assignable.

Assuming additive model and normal distribution we can analyze the experiment as a one-way classification.

Suppose after conducting the experiment we get the following table of values.

| Varieties | Observations |
|---|---|
| $V_1$ | $y_{11}$ $y_{12}$ $\cdots$ $y_{1n_1}$ |
| $V_2$ | $y_{21}$ $y_{22}$ $\cdots$ $y_{2n_2}$ |
| … … | … … … … … |
| $V_k$ | $y_{k1}$ $y_{k2}$ $\cdots$ $y_{kn_k}$ |

The $n_i$ observations corresponding to the i-th variety differ only due to random error. But the means of the i-th and j-th varieties will differ due to the difference in effects of $V_i$ and $V_j$. Let $\alpha_1, \ldots, \alpha_k$ be the effects of the varieties $V_1, \ldots, V_k$. If the test of the hypothesis $H_0: \alpha_1 = \cdots = \alpha_k$ is not significant we may assume that the variety effects are uniform and in this case no special choice is to be made based on the quality of seeds.

On the other hand if the test is significant we cannot assume that the variety effects are the same. In this case we test individual hypotheses of the form $\alpha_i = \alpha_j$. If the corresponding test is found significant and if we see that the mean yield due to the i-th variety (i.e., $\bar{y}_{i.}$), is greater than the mean yield due to the j-th variety (i.e., $\bar{y}_{j.}$), then without any hesitation we can say that the i-th variety is superior to the j-th variety.

### 3.7 Two-Way Classification with a Single Observation per Cell

Let there be pq independent observations of the variable y, each of which can be specified in terms of the categories of two classes A and B. Let $A_1, \dots, A_p$ be the p categories in the A-classification and $B_1, \dots, B_q$ be the categories in the B-classification. The observations may be arranged in the following tabular form:

| A \ B | $B_1$ $B_2$ … $B_q$ | Total |
|---|---|---|
| $A_1$ | $y_{11}$ $y_{12}$ … $y_{1q}$ | $y_{1.}$ |
| $A_2$ | $y_{21}$ $y_{22}$ … $y_{2q}$ | $y_{2.}$ |
| … | … … … | … |
| $A_p$ | $y_{p1}$ $y_{p2}$ … $y_{pq}$ | $y_{p.}$ |
| Total | $y_{.1}$ $y_{.2}$ … $y_{.q}$ | $y_{..}$ |

Let $\alpha_1, \dots, \alpha_p$ be the effects due to the categories $A_1, \dots, A_p$ and $\beta_1, \dots, \beta_q$ the effects due to the categories $B_1, \dots, B_q$. Assuming that the effects are additive, the model will be

$$y_{ij} = \alpha_i + \beta_j + \varepsilon_{ij}$$

where $\varepsilon_{ij}$ is a random error which we assume to be distributed normally around zero with a common variance $\sigma^2$. The two hypotheses which may be tested from this data are

$$H_1\colon \alpha_1 = \cdots = \alpha_p (= \alpha, \text{say})$$

$$H_2\colon \beta_1 = \cdots = \beta_q (= \beta, \text{say}).$$

Let

$$y = (y_{11}\, y_{12} \cdots y_{1q}\, y_{21}\, y_{22} \cdots y_{2q} \cdots y_{p1}\, y_{p2} \cdots y_{pq})$$

and

$$\theta = (\alpha_1, \dots, \alpha_p; \beta_1, \dots, \beta_q).$$

Then the observational equations become

$$E(y) = \theta A^T,$$

where

$$A^T = \begin{bmatrix} J & o & \dots \\ o & J & o \\ \dots & \dots & J \\ I_q & I_q & I_q \end{bmatrix}$$

where $J = (1, 1, \dots, 1)$ is $1 \times q$, $o = (0, 0, \dots, 0)$ is $1 \times q$ and $I_q$ is an identity matrix of order q. Obviously Rank $A = (p + q - 1)$ so that not all parametric functions are estimable. Consider a linear parametric function of the form

$$c\theta^T = c_1\alpha_1 + \dots + c_p\alpha_p + c_1'\beta_1 + \dots + c_q'\beta_q.$$

Then the necessary and sufficient condition for the estimability of $c\theta^T$ is that the vector c should be a linear combination of the rows of A. In other words $c^T$ should be a linear combination of the rows of $A^T$. But in $A^T$ the sum of the first p rows is equal to the sum of the next q rows. Therefore in $c^T$ the sum of the first p elements should equal the sum of the last q elements. Hence any linear parametric function $c\theta^T$ is estimable if and only if it is such that

$$c_1 + \dots + c_p = c_1' + \dots + c_q'.$$

Therefore we see that individual parameters are not estimable, whereas contrasts, with sum of coefficients zero, are estimable. In particular "elementary contrasts" of the form $\alpha_i - \alpha_j, \beta_i - \beta_j$, etc.( in which we are interested) are estimable.

The residual S.S. is obtained by minimizing

$$L = \sum_{i=1}^{p}\sum_{j=1}^{q}(y_{ij} - \alpha_i - \beta_j)^2$$

w.r.t. the $\alpha'$s and $\beta'$s. The minimizing equations are $\frac{\partial L}{\partial \alpha_i} = 0, i = 1, \dots, p$ and $\frac{\partial L}{\partial \beta_j} = 0, j = 1, \dots, q$. These give

$$\sum_{j=1}^{q}(y_{ij} - \alpha_i - \beta_j) = 0$$

and

$$\sum_{i=1}^{p}(y_{ij} - \alpha_i - \beta_j) = 0.$$

Thus the normal equations are

$$y_{i.} - q\alpha_i - \beta_. = 0, \quad i = 1, \ldots p \tag{1}$$

$$\text{and } y_{.j} - \alpha_. - p\beta_j = 0, \quad j = 1, \ldots, q \tag{2}$$

where $\alpha_. = \sum_i \alpha_i$ and $\beta_. = \sum_j \beta_j$.

Out of the $(p+q)$ normal equations only $(p+q-1)$ are independent, since $\sum_i(1) = \sum_j(2)$, and so we may impose one condition to solve them. Let us impose the condition $\beta_. = 0$. Then, from (1), we get

$$\hat{\alpha}_i = \frac{y_{i.}}{q} = \bar{y}_{i.},$$

where $\bar{y}_{i.}$ is the mean of the observations in the i-th row. Therefore

$$\hat{\alpha}_. = \sum_i \hat{\alpha}_i = \sum_i \frac{y_{i.}}{q} = \frac{y_{..}}{q},$$

where $y_{..}$ is the grand total of all the observations. From the second equation we now have

$$\hat{\beta}_j = \frac{y_{.j} - \frac{y_{..}}{q}}{p} = \bar{y}_{.j} - \bar{y}_{..}\,.$$

where $\bar{y}_{.j}$ is the mean of the observations in the j-th column and $\bar{y}_{..}$ is the grand mean.

The unconditional minimum now is

$$S^2 = \sum_{i,j}^{q} \left(y_{ij} - \hat{\alpha}_i - \hat{\beta}_j\right)^2$$

$$= \sum_{i,j}^{q} \left(y_{ij} - \hat{\alpha}_i - \hat{\beta}_j\right) y_{ij}, \quad \text{by the normal equations}$$

$$= \sum_{i,j}^{q} \left(y_{ij} - \bar{y}_{i.} - \bar{y}_{.j} + \bar{y}_{..}\right) y_{ij}$$

$$= \sum_{i,j} y_{ij}^2 - \sum_{i,j} \bar{y}_{i.} y_{ij} - \sum_{i,j} \bar{y}_{.j} y_{ij} + \sum_{i,j} \bar{y}_{..} y_{ij}$$

$$= \sum_{i,j} y_{ij}^2 - \sum_i \frac{y_{i.}^2}{q} - \sum_j \frac{y_{.j}^2}{p} + \frac{y_{..}^2}{pq}, \tag{3}$$

and the d. f. of $S^2$ will be $pq - (p + q - 1)$.

**(i) Consider the hypotheses $H_1$**

This is a linear hypothesis with $(p - 1)$ d · f ·. The error sum of squares under $H_1$ is

$$L_1 = \sum_{i,j} (y_{ij} - \alpha - \beta_j)^2$$

and the minimizing equations are

$$\sum_{i,j} (y_{ij} - \alpha - \beta_j) = 0, \quad \sum_i (y_{ij} - \alpha - \beta_j) = 0.$$

$$\text{i.e., } y_{..} - pq\alpha - p\beta. = 0 \tag{4}$$

$$\text{and } \; y_{.j} - p(\alpha + \beta_j) = 0 \tag{5}$$

From (5), the least square estimate of $(\alpha + \beta_j)$ is

$$\widehat{(\alpha + \beta_j)} = \frac{y_{.j}}{p}.$$

Therefore the residual sum of squares under $H_1$ is

$$S_1^2 = \sum_{i,j} \left(y_{ij} - \frac{y_{.j}}{p}\right)^2$$

$$= \sum_{i,j} \left(y_{ij} - \frac{y_{.j}}{p}\right) y_{ij}, \text{ by the normal equations}$$

$$= \sum_{i,j} y_{ij}^2 - \sum_j \frac{y_{.j}^2}{p}$$

and this has $pq - (p + q - 1) + p - 1 = q(p - 1)$ d.f.. Therefore

S.S. due to deviation in $\alpha$-effects is

$$S_1^2 - S^2 = \sum_i \frac{y_{i.}^2}{q} - \frac{y_{..}^2}{pq}, \tag{6}$$

which has $(p-1)$ d.f..

Under assumption of normality the F-ratio to test $H_1$ is given by

$$F(p-1, pq-(p+q-1)) = \frac{(S_1^2 - S^2)/(p-1)}{S^2/(pq-(p+q-1))}.$$

**(ii) Consider $H_2$**

This is a linear hypothesis with $(q-1)$ d.f.. The residual S.S. under $H_2$ is obtained by minimizing

$$L_2 = \sum_{i,j} (y_{ij} - \alpha_i - \beta)^2.$$

The normal equations are

$$\sum_j (y_{ij} - \alpha_i - \beta) = 0$$

$$\sum_{i,j} (y_{ij} - \alpha_i - \beta)^2 = 0.$$

Considering the first equation we get the least square estimate of $\alpha_i + \beta$ to be

$$(\widehat{\alpha_i + \beta}) = \frac{y_{i.}}{q}.$$

Therefore the residual S.S. under $H_2$ is

$$S_2^2 = \sum_{i,j} \left(y_{ij} - \frac{y_{i.}}{q}\right)^2$$

$$= \sum_{i,j} \left(y_{ij} - \frac{y_{i.}}{q}\right) y_{ij}, \text{ by the normal equation}$$

$$= \sum_{i,j} y_{ij}^2 - \sum_j \frac{y_{i.}^2}{q}.$$

This has $pq - (p+q-1) + q - 1 = p(q-1)$ d.f..

The F-ratio in this case is

$$F\big(q-1, pq-(p+q-1)\big) = \frac{(S_2^2 - S^2)/(q-1)}{S^2/(pq-(p+q-1))}.$$

Let us proceed to set up the analysis of variance table by splitting the total sum of squares into sum of squares due to the A-classification, sum of squares due to the B-classification and the error sum of squares. The total S.S. is

$$\sum_{i,j} (y_{ij} - \bar{y}_{..})^2 = \sum_{i,j} y_{ij}^2 - \frac{y_{..}^2}{pq}.$$

The splitting of this total S.S. is as shown below:

$$\sum_{i,j} y_{ij}^2 - \frac{\bar{y}_{..}^2}{pq} = \sum_{i,j} (y_{ij} - \bar{y}_{..})^2$$

$$= \sum_{i,j} [(y_{ij} - \bar{y}_{i.} - \bar{y}_{.j} - \bar{y}_{..}) + (\bar{y}_{i.} - \bar{y}_{..}) + (\bar{y}_{.j} - \bar{y}_{..})]^2$$

$$= \sum_{i,j} (y_{ij} - \bar{y}_{i.} - \bar{y}_{.j} - \bar{y}_{..})^2 + \sum_{i,j} (\bar{y}_{i.} - \bar{y}_{..})^2$$

$$+ \sum_{i,j} (\bar{y}_{.j} - \bar{y}_{..})^2$$

$$+2\sum_{i,j} (y_{ij} - \bar{y}_{i.} - \bar{y}_{.j} - \bar{y}_{..})(\bar{y}_{i.} - \bar{y}_{..})$$

$$+2\sum_{i,j} (y_{ij} - \bar{y}_{i.} - \bar{y}_{.j} - \bar{y}_{..})(\bar{y}_{.j} - \bar{y}_{..})$$

$$+2\sum_{i,j} (\bar{y}_{i.} - \bar{y}_{..})(\bar{y}_{.j} - \bar{y}_{..}).$$

But

$$\sum_{i,j} (y_{ij} - \bar{y}_{i.} - \bar{y}_{.j} - \bar{y}_{..})(\bar{y}_{i.} - \bar{y}_{..})$$

$$= \sum_{i} (\bar{y}_{i.} - \bar{y}_{..}) \sum_{j} (y_{ij} - \bar{y}_{i.} - \bar{y}_{.j} - \bar{y}_{..})$$

$$= \sum_{i} (\bar{y}_{i.} - \bar{y}_{..})(y_{i.} - q\bar{y}_{i.} - q\bar{y}_{..} + q\bar{y}_{..})$$

$$= 0$$

Similarly, the other cross-products also vanish. Therefore

$$\sum_{i,j} y_{ij}^2 - \frac{\bar{y}_{..}^2}{pq} = \sum_{i,j} (y_{ij} - \bar{y}_{i.} - \bar{y}_{.j} - \bar{y}_{..})^2 + \sum_{i,j} (\bar{y}_{i.} - \bar{y}_{..})^2 + \sum_{i,j} (\bar{y}_{.j} - \bar{y}_{..})^2.$$

The first term is $S^2$, the residual S.S. obtained in (1). Again,

$$\sum_{i,j} (\bar{y}_{i.} - \bar{y}_{..})^2 = q \sum_{i} (\bar{y}_{i.} - \bar{y}_{..})^2$$
$$= q \sum_{i} \left( \frac{y_{i.}}{q} - \frac{y_{..}}{pq} \right)^2$$
$$= \sum_{i} \frac{y_{i.}^2}{q} - \frac{y_{..}^2}{pq},$$

which is the S.S. due to the deviation in A-effects obtained in (4). Similarly,

$$\sum_{i,j} (\bar{y}_{.j} - \bar{y}_{..})^2 = \sum_{j} \frac{y_{.j}^2}{p} - \frac{y_{..}^2}{pq}, \tag{7}$$

is the S.S. due to the deviation in B-effects.

The analysis of variance table is as given below.

| Variation due to | S.S. | d.f. | M.S.S. | F-ratio |
|---|---|---|---|---|
| A-classification | $\sum_{i} \frac{y_{i.}^2}{q} - \frac{y_{..}^2}{pq}$ | $p - 1$ | $V_A$ | $\frac{V_A}{V_E}$ |
| B-classification | $\sum_{j} \frac{y_{.j}^2}{p} - \frac{y_{..}^2}{pq}$ | $q - 1$ | $V_B$ | $\frac{V_B}{V_E}$ |
| Residual | (By sub) | (By sub) | $V_E$ | |
| Total | $\sum_{i,j} y_{ij}^2 - \frac{y_{..}^2}{pq}$ | $pq - 1$ | | |

## 3.8 Testing of Individual Hypotheses

In case $\frac{V_A}{V_E}$ is significant, $H_1$ is no longer true and we proceed to test individual hypotheses such as $\alpha_i = \alpha_j$ using the t-test.

$$t = \frac{\hat{\alpha}_i - \hat{\alpha}_j}{\sqrt{\text{unbiased estimate of } V(\hat{\alpha}_i - \hat{\alpha}_j)}}.$$

We have seen that contrasts of the form $\alpha_i - \alpha_j$ are estimable and so are uniquely estimated by the Principle of Substitution (Gauss-Markov Theorem) and hence the unique estimate of $\alpha_i - \alpha_j$ is

$$\hat{\alpha}_i - \hat{\alpha}_j = \bar{y}_{i.} - \bar{y}_{j.} = \frac{1}{q}(y_{i.} - y_{j.}).$$

Next, let us get the variance of $\hat{\alpha}_i - \hat{\alpha}_j$. We know that if $c\theta^T$ is estimable, then the unique best unbiased estimate of $c\theta^T$ is $\lambda A^T y^T$ where $\lambda$ is any solution of $\lambda A^T A = c$ and the variance of the estimator is

$$V(\lambda A^T y^T) = c\lambda^T \sigma^2.$$

Now $yA = (y_{1.}, y_{2.}, \ldots, y_{p.}; y_{.1}, y_{.2}, \ldots, y_{.q})$. Let $\lambda = (\lambda_1, \ldots, \lambda_p;\ \lambda_1', \ldots,$ $\lambda_q')$. If we take $c_i = 1, c_j = -1$ and all other elements zero in c, then $c\theta^T = \alpha_i - \alpha_j$, the contrast whose least square estimate is $\frac{(y_{i.} - y_{j.})}{q}$, which is unique by Gauss-Markov Theorem.

$$\text{i.e.,}\ \lambda A^T y^T = \frac{1}{q}(y_{i.} - y_{j.}).$$

So we have to take $\lambda_i = \frac{1}{q}, \lambda_j = -\frac{1}{q}$ and all other elements zero in $\lambda$ and hence the variance of the estimator will be

$$c\lambda^T \sigma^2 = \left(\frac{1}{q} + \frac{1}{q}\right)\sigma^2 = \frac{2}{q}\sigma^2.$$

$$\text{i.e.,}\ V(\hat{\alpha}_i - \hat{\alpha}_j) = \frac{2}{q}\sigma^2$$

and an unbiased estimator of $\sigma^2$ is $\frac{S^2}{n-r}$ where n is the total number of observations and r is the rank of the observation matrix. In this case $n = pq$ and $r = p + q - 1$. So an unbiased estimate of $V(\hat{\alpha}_i - \hat{\alpha}_j)$ is

$$\left(\frac{2}{q}\right)\frac{S^2}{pq - (p + q - 1)} = \frac{2}{q}V_E$$

And the t-criterion becomes

$$t = \frac{\bar{y}_{i.} - \bar{y}_{j.}}{\sqrt{\frac{2}{q}\frac{S^2}{(p-1)(q-1)}}} = \frac{\bar{y}_{i.} - \bar{y}_{j.}}{\sqrt{\frac{2}{q}V_E}}$$

which, under the assumption of normality, is a Student t with $(p - 1)(q - 1)$ d.f..

Again, if $\frac{V_B}{V_E}$ is found significant, we can test individual hypotheses of the from $\beta_i = \beta_j$, and by a similar procedure as above, the test criterion will be found to be

$$t = \frac{\bar{y}_{.i} - \bar{y}_{.j}}{\sqrt{\frac{2}{q}V_E}}$$

which is a Student t with $(p - 1)(q - 1)$ d.f..

Note:

In the above two tests, one fundamental assumption regarding the t- has been violated, because

$$\frac{\hat{\alpha}_i - \hat{\alpha}_j}{\sqrt{\text{unbiased estimate of } V(\hat{\alpha}_i - \hat{\alpha}_j)}} \text{ or } \frac{\hat{\beta}_i - \hat{\beta}_j}{\sqrt{\text{unbiased estimate of } V(\hat{\beta}_i - \hat{\beta}_j)}}$$

will be a Student t if and only if $\alpha_i$ and $\alpha_j$ or $\beta_i$ and $\beta_j$ are independently estimable and here they are not, but only the elementary contrasts of the form $\alpha_i - \alpha_j$ or $\beta_i - \beta_j$ are estimable. However, in the absence of more reliable tests, the above t-test may be assumed valid, but it is better to bear in mind the following points when applying the t-test:

i) See whether the means are randomly selected.

ii) It is better to apply t-test if the test is preconceived when we design the experiment.

iii) Apply the t-test only if the F-test is significant even at the 10% level.

## 3.9 Orthogonal Effects

Two effects are said to be orthogonal if they can be estimated independently. So far we have been considering only orthogonal effects. In that case the column corresponding to d. f. in the Analysis of Variance table is additive and the d. f. of the residual S.S. is obtained by subtraction. This method is applicable only if the effects are orthogonal. In two-way classification, we are concerned with orthogonal effects.

## 3.10 Two-Way Classification with Multiple but Equal Number of Observations in Cells

Let there be npq independents observations classified according to p A-classifications and q B-classifications so that each cell contains n observations. Let the k-th observation in the (i,j)-th cell be $y_{ijk}$.

### Interaction

If the B-classification differs for fixed A-classification or vice versa it is said to be due to interaction. In other words interaction signifies the combined effect of the A and B-classifications.

Let $\alpha_1, \ldots, \alpha_p$ be the row effects and $\beta_1, \ldots, \beta_q$ the column effects. Let $\gamma_{ij}$ be the combined effect (i.e., interaction) of $A_i$ and $B_j$. Assuming additivity of effects, each observation may be written as

$$y_{ijk} = \alpha_i + \beta_j + \gamma_{ij} + \varepsilon_{ijk},$$

where $\varepsilon_{ijk}$ is a chance variation which we assume to be normally distributed around zero with a common but unknown variance $\sigma^2$. Then the observational equations will be

$$E(y_{ijk}) = \mu_{ij},$$

where $\mu_{ij} = \alpha_i + \beta_j + \gamma_{ij}$, where there is interaction and $\mu_{ij} = \alpha_i + \beta_j$, when there is no interaction.

The first hypothesis we want to test is

$$\mu_{ij} = \alpha_i + \beta_j,$$

i.e., whether there is interaction or not. The procedure for further tests will differ according as this first test is significant or not.

Notations

$$\sum_i y_{ijk} = y_{.jk}, \quad \sum_j y_{ijk} = y_{i.k}, \quad \sum_k y_{ijk} = y_{ij.},$$

$$\sum_{i,j} y_{ijk} = y_{..k}, \quad \sum_{i,k} y_{ijk} = y_{.j.}, \quad \sum_{j,k} y_{ijk} = y_{i..},$$

$$\sum_{i,j,k} y_{ijk} = y_{...},$$

$$\sum_i \alpha_i = \alpha_{.}, \quad \sum_j \beta_j = \beta_{.}, \quad \sum_{i,j} \gamma_{ij} = \gamma_{..}, \quad \sum_i \gamma_{ij} = \gamma_{.j} \text{ and } \sum_j \gamma_{ij} = \gamma_{i.}\,.$$

Treating the $\mu_{ij}$ as free parameters, the residual S.S. when there is interaction is gotten by minimizing, w.r.t. $\mu_{ij}$,

$$L = \sum_{i,j,k} (y_{ijk} - \mu_{ij})^2,$$

where $\mu_{ij} = \alpha_i + \beta_j + \gamma_{ij}$. The minimizing equation is $\frac{\partial L}{\partial \mu_{ij}} = 0$.

$$\text{i.e.,} \sum_k (y_{ijk} - \mu_{ij}) = 0\,.$$

which gives the least square estimate of $\mu_{ij}$ as $\hat{\mu}_{ij} = \frac{y_{ij.}}{n}$.

Therefore the minimum error S.S. is

$$\begin{aligned} S^2 &= \sum_{i,j,k} \left(y_{ijk} - \frac{y_{ij.}}{n}\right) y_{ijk} \\ &= \sum_{i,j,k} y_{ijk}^2 - \sum_{i,j} \frac{y_{ij.}^2}{n} \\ &= \left(\sum_{i,j,k} y_{ijk}^2 - \frac{y_{...}^2}{npq}\right) - \left(\sum_{i,j} \frac{y_{ij.}^2}{n} - \frac{y_{...}^2}{npq}\right), \end{aligned}$$

which has $npq - pq = (n-1)pq$ d.f..

Assuming $H_0: \gamma_{ij} = 0$ for all i and j (which has $(p-1)(q-1)$ d.f.), the total error S.S. is

$$L_0 = \sum_{i,j,k} (y_{ijk} - \alpha_i - \beta_j)^2$$

which has to be minimized w.r.t. the α's and β's. The conditions are

$$\frac{\partial L}{\partial \alpha_i} = 0, \quad \frac{\partial L}{\partial \beta_j} = 0$$

giving the normal equations as

$$\sum_{j,k} (y_{ijk} - \alpha_i - \beta_j) = 0,$$

$$\sum_{i,k} (y_{ijk} - \alpha_i - \beta_j) = 0,$$

$$\text{i.e.,} \, y_{i..} - nq\alpha_i - n\beta_{.} = 0, \qquad (1)$$

$$\text{and} \quad y_{.j.} - n\alpha_{.} - np\beta_j = 0, \qquad (2)$$

$\Sigma_i(1) = \Sigma_j(2)$ so that of the $(p + q)$ equations only $(p + q - 1)$ are independent. Therefore we may impose one condition to solve them. Let us impose the condition $\beta_{.} = 0$. Then the first equation gives an estimate of $\alpha_i$ to be

$$\hat{\alpha}_i = \frac{y_{i..}}{nq}, i = 1,2, \dots, p.$$

$$\hat{\alpha}_{.} = \sum_i \hat{\alpha}_i = \frac{y_{...}}{nq}$$

and so from the second equation an estimate of $\beta_j$ is

$$\hat{\beta}_j = \frac{y_{.j.}}{np} - \frac{y_{...}}{npq}.$$

Therefore the conditional minimum under $H_0$ is

$$S_0^2 = \sum_{i,j,k} (y_{ijk} - \frac{y_{i..}}{nq} - \frac{y_{.j.}}{np} + \frac{y_{...}}{npq}) y_{ijk}$$

$$= \left( \sum_{i,j,k} y_{ijk}^2 - \frac{y_{...}^2}{npq} \right) - \left( \sum_i \frac{y_{i..}^2}{nq} - \frac{y_{...}^2}{npq} \right) - \left( \sum_j \frac{y_{.j.}^2}{np} - \frac{y_{...}^2}{npq} \right)$$

the d.f. being $(n - 1)pq + (p - 1)(q - 1) = npq - p - q + 1$. Therefore the S.S. due to interaction is

$$S_0^2 - S^2 = \left(\sum_{i,j} \frac{y_{ij.}^2}{n} - \frac{y_{...}^2}{npq}\right) - \left(\sum_i \frac{y_{i..}^2}{nq} - \frac{y_{...}^2}{npq}\right)$$
$$- \left(\sum_j \frac{y_{.j.}^2}{np} - \frac{y_{...}^2}{npq}\right)$$

which has $(p-1)(q-1)$ d.f.

Under the assumption of normality

$$\frac{(S_0^2 - S^2)/(p-1)(q-1)}{S^2/(n-1)pq}$$

is an F with $(p-1)(q-1)$ and $(n-1)pq$ d.f.. If F is significant, there is interaction. When F is not significant, we may assume $\mu_{ij} = \alpha_i + \beta_j$, and in this case two further hypotheses which may be tested are

$$H_1: \alpha_1 = \alpha_2 = \cdots = \alpha_p (= \alpha, \text{say}),$$

$$H_2: \beta_1 = \beta_2 = \cdots = \beta_q (= \beta, \text{say}).$$

The unconditional minimum in both cases is $S_0^2$. Under $H_1$, we have

$$E(y_{ijk}) = \mu_{ij} \text{ where } \mu_{ij} = \alpha + \beta_j,$$

so that the total error S.S. is

$$L = \sum_{i,j,k} (y_{ijk} - \alpha - \beta_j)^2.$$

Minimizing L w.r.t. $\alpha$ and $\beta_j$'s, the least square estimate of $\alpha + \beta_j$ is

$$\widehat{(\alpha + \beta_j)} = \frac{y_{.j.}}{np},$$

and the conditional minimum under $H_1$ is

$$S_1^2 = \sum_{i,j,k} (y_{ijk} - \frac{y_{.j.}}{np}) y_{ijk}$$
$$= \left(\sum_{i,j,k} y_{ijk}^2 - \frac{y_{...}^2}{npq}\right) - \left(\sum_j \frac{y_{.j.}^2}{np} - \frac{y_{...}^2}{npq}\right).$$

Therefore the S.S. due to the A-classification is

$$S_1^2 - S_0^2 = \sum_i \frac{y_{i..}^2}{nq} - \frac{y_{...}^2}{npq}$$

with d.f. $(p-1)$. Similarly, the S.S. due to the B-classification is

$$\sum_j \frac{y_{.j.}^2}{np} - \frac{y_{...}^2}{npq}$$

with d.f. $(q-1)$.

Now, let us proceed to set up the analysis of variance table. For that we have to split the total sum of squares

$$T = \sum_{i,j,k} \left( y_{ijk} - \frac{y_{...}}{npq} \right)^2$$

$$= \sum_{i,j,k} y_{ijk}^2 - \frac{y_{...}^2}{npq}$$

with d.f. $(npq-1)$ into its various components. We have

$$T = \sum_{i,j,k} (y_{ijk} - \frac{y_{...}}{npq})^2$$

$$= \sum_{i,j,k} (y_{ijk} - \frac{y_{ij.}}{n} + \frac{y_{ij.}}{n} - \frac{y_{...}}{npq})^2$$

$$= \sum_{i,j,k} (y_{ijk} - \frac{y_{ij.}}{n})^2 + \sum_{i,j,k} (\frac{y_{ij.}}{n} - \frac{y_{...}}{npq})^2,$$

$$\text{(since the cross} - \text{product term vanishes.)}$$

$$= \left( \sum_{i,j,k} y_{ijk}^2 - \sum_{i,j} \frac{y_{ij.}^2}{n} \right) - \left( \sum_{i,j} \frac{y_{ij.}^2}{n} - \frac{y_{...}^2}{npq} \right).$$

The first term on the right gives the S.S. due to the within cell variation or residual and the second term gives the S.S. due to the between cell variation with respective d.f. $(n-1)pq$ and $(pq-1)$.

The between cell variation can be further split into its components, viz. S.S. due to the A-classification, S.S. due to the B-classification and S.S. due to interaction. This is done as follows.

The between cell S.S. is

$$B = \sum_{i,j,k} (\frac{y_{ij.}}{n} - \frac{y_{...}}{npq})^2$$

$$= \sum_{i,j,k} [\left(\frac{y_{ij.}}{n} - \frac{y_{i..}}{nq} - \frac{y_{.j.}}{np} + \frac{y_{...}}{npq}\right) + \left(\frac{y_{i..}}{nq} - \frac{y_{...}}{npq}\right) + (\frac{y_{.j.}}{np} - \frac{y_{...}}{npq})]^2$$

$$= \sum_{i,j,k} \left(\frac{y_{ij.}}{n} - \frac{y_{i..}}{nq} - \frac{y_{.j.}}{np} + \frac{y_{...}}{npq}\right)^2 + \sum_{i,j,k} \left(\frac{y_{i..}}{nq} - \frac{y_{...}}{npq}\right)^2$$

$$+ \sum_{i,j,k} (\frac{y_{.j.}}{np} - \frac{y_{...}}{npq})^2$$

$$\text{(since the cross – product terms all vanish.)}$$

Now,

$$\sum_{i,j,k} \left(\frac{y_{i..}}{nq} - \frac{y_{...}}{npq}\right)^2 = \sum_{i} \frac{y_{i..}^2}{nq} - \frac{y_{...}^2}{npq}.$$

This is the S.S. due to the A-effects (when the interaction is zero.) Again,

$$\sum_{i,j,k} \left(\frac{y_{.j.}}{np} - \frac{y_{...}}{npq}\right)^2 = \sum_{j} \frac{y_{.j.}^2}{np} - \frac{y_{...}^2}{npq}.$$

This is the S.S. due to the B-effects (when the interaction is zero.) Finally,

$$\sum_{i,j,k} \left(\frac{y_{ij.}}{n} - \frac{y_{i..}}{nq} - \frac{y_{.j.}}{np} + \frac{y_{...}}{npq}\right)^2$$

$$= \left(\sum_{i,j} \frac{y_{ij.}^2}{n} - \frac{y_{...}^2}{npq}\right) - \left(\sum_{i} \frac{y_{i..}^2}{nq} - \frac{y_{...}^2}{npq}\right) - \left(\sum_{j} \frac{y_{.j.}^2}{np} - \frac{y_{...}^2}{npq}\right),$$

which is the S.S. due to interaction. Therefore

$$\text{Between cell S. S.}$$
$$= \text{S. S. due to A} + \text{S. S. due to B}$$
$$+ \text{S. S. due to interaction.}$$

Correspondingly we have a splitting of the $\text{d. f.}$ viz, ,

$$pq - 1 = (p - 1) + (q - 1) + (p - 1)(q - 1).$$

The analysis of variance is set out in the following table.

| Variation due to | S.S. | d.f. | M.S.S | F-ratio |
|---|---|---|---|---|
| A classification (A) | $\sum_i \frac{y_{i..}^2}{nq} - \frac{y_{...}^2}{npq}$ | $p-1$ | $V_A$ | |
| B classification (B′) | $\sum_j \frac{y_{.j.}^2}{np} - \frac{y_{...}^2}{npq}$ | $q-1$ | $V_{B'}$ | |
| Interaction (A × B) | $(B - A - B')$ | $(p-1)(q-1)$ | $V_{AB}$ | $\frac{V_{AB}}{V_E}$ |
| Between cell (B) | $\sum_{i,j} \frac{y_{ij.}^2}{n} - \frac{y_{...}^2}{npq}$ | $pq-1$ | | |
| Within cell or Residual ($S^2$) | $T - B$ | $(n-1)pq$ (By sub) | $V_E$ | |
| Total | $\sum_{i,j,k} y_{ijk}^2 - \frac{y_{...}^2}{npq}$ | $npq-1$ | | |

First we test for interaction using the F-ratio $V_{AB}/V_E$. Two cases arise.

**Case (i):**

Interaction is not significant. We then proceed to test the A and B effects. For this we test $V_A$ and $V_{B'}$ against the interaction mean square $V_{AB}$ or the residual mean square $V_E$, whichever is greater. This is to guard against any bias due to small effects of interaction which could not be detected by the test which is indicated when the interaction mean square exceeds the error. When both the mean squares are of the same magnitude, a common estimate can be obtained by taking a weighted average of the two with corresponding d. f. as weights. If this new error mean square is $V_{E'}$, we test $V_A$ and $V_{B'}$ against $V_{E'}$.

**Note:**

In the case when the interaction is not significant, there is no use in considering the term $\gamma_{ij}$. The whole analysis reduces to the ordinary two-way classification, the only difference being that we take the sum of the n observations in each cell as the observation in that cell.

**Case (ii):**

If the interaction is significant further analysis becomes a bit complicated. Differences in A-classification have to be tested for each B class and vice versa. Suppose we want to test the hypothesis $\alpha_1 = \alpha_2 = \cdots = \alpha_p$. Since there is significant interaction we have to test this hypothesis for every B class. Let us specify a B-class, say $B_j$. Corresponding to each A-class we get n observations. So this reduces to the one-way classification. We are getting p samples of size n each. The hypothesis can be tested exactly of size n each. The hypothesis can be tested exactly as in the case of one-way classification. In the case of the hypothesis $\beta_1 = \beta_2 = \cdots = \beta_q$, we specify any A-class and proceed as before.

## Computational Procedure

Using the given two-way table we form the following table from which all the required quantities can be computed.

| B / A | $B_1$ | $B_2$ | ... | $B_q$ | Sum |
|---|---|---|---|---|---|
| $A_1$ | $y_{11.}$ | $y_{12.}$ | ... | $y_{1q.}$ | $y_{1..}$ |
| $A_2$ | $y_{21.}$ | $y_{22.}$ | ... | $y_{2q.}$ | $y_{2..}$ |
| ... | .. | ... | ... | ... | ... |
| $A_p$ | $y_{p1.}$ | $y_{p2.}$ | ... | $y_{pq.}$ | $y_{p..}$ |
| Sum | $y_{.1.}$ | $y_{.2.}$ | ... | $y_{.q.}$ | $y_{...}$ |

## 3.11 Two-Way Classification with Multiple and Unequal Number of Observations in Cells, but There is at Least One Observation in Each Cell

Let there be $n_{ij}$ observations in the (i,j)-th cell, $i = 1, \ldots, p, j = 1, \ldots, q$. The A-classification we shall call blocks and the B-classification treatments. There are p blocks and q treatments. Let $\alpha_1, \alpha_2, \ldots, \alpha_p$ be the block effects and $\beta_1, \beta_2, \dots, \beta_q$ the treatment effects. Let $y_{ijk}$ be the k-th observation in the (i,j)-th cell, $k = 1,2,\dots,n_{ij}$. The model assumed is

$$y_{ijk} = \alpha_i + \beta_j + \gamma_{ij} + \varepsilon_{ijk}$$

where $\gamma_{ij}$ is the interaction (i.e., the combined effects of the blocks and treatments) and $\varepsilon_{ijk}$ is a random error which, as usual, we assume to be distributed normally around zero with a common but unknown variance $\sigma^2$. Therefore the observational equations become

$$E(y_{ijk}) = \mu_{ij},$$

where $\mu_{ij} = \alpha_i + \beta_j + \gamma_{ij}$.

Let $H_0: \gamma_{ij} = 0$. The total error S.S. is

$$L = \sum_{i,j,k} (y_{ijk} - \mu_{ij})^2$$

where $\mu_{ij} = \alpha_i + \beta_j + \gamma_{ij}$. Considering $\mu_{ij}$ as an independent parameter the minimizing condition is

$$\frac{\partial L}{\partial \mu_{ij}} = 0.$$

$$\text{i.e.,} \sum_k (y_{ijk} - \mu_{ij}) = 0,$$

where $\hat{\mu}_{ij} = \frac{y_{ij.}}{n_{ij}} = \bar{y}_{ij.}$. Therefore the unconditional minimum is

$$S^2 = \sum_{i,j,k} (y_{ijk} - \frac{y_{ij.}}{n_{ij}}) y_{ijk}$$

$$= \left( \sum_{i,j,k} y_{ijk}^2 - \frac{y_{...}^2}{n_{..}} \right) - \left( \sum_{i,j} \frac{y_{ij.}^2}{n_{ij}} - \frac{y_{...}^2}{n_{..}} \right),$$

where $n_{..} = \sum_{i,j} n_{ij}$. The first term in brackets is the total S.S. and the second is the between cell S.S.. The d.f. of $S^2$ is $\sum_{i,j} n_{ij} - pq = n_{..} - pq$.

Now $H_0$ is a linear hypothesis with $(p-1)(q-1)$ d.f. Under $H_0$, the error S.S. is

$$L_0 = \sum_{i,j,k} (y_{ijk} - \alpha_i - \beta_j)^2 .$$

The minimizing conditions are $\frac{\partial L_0}{\partial \alpha_i} = 0, \frac{\partial L_0}{\partial \beta_j} = 0$. These give the normal equations

$$y_{i..} - n_{i.}\alpha_i - \sum_j n_{ij}\,\beta_j = 0 \qquad (1)$$

$$\text{and } y_{.j.} - \sum_i n_{ij}\alpha_i - n_{.j}\beta_j = 0 \qquad (2)$$

since $\sum_i(1) = \sum_j(2)$ we have only $(p + q - 1)$ independent normal equations to estimate the $(p + q)$ unknown parameters. From (ii),

$$\beta_j = \frac{y_{.j.}}{n_{.j}} - \frac{\sum_i n_{ij}\alpha_i}{n_{.j}} = \frac{y_{.j.}}{n_{.j}} - \frac{\sum_l n_{lj}\alpha_l}{n_{.j}}.$$

Eliminating $\beta_j$ from the first equation, we have

$$y_{i..} - n_{i.}\alpha_i - \sum_j n_{ij}\left\{\frac{y_{.j.}}{n_{.j}} - \frac{\sum_l n_{lj}\alpha_l}{n_{.j}}\right\} = 0.$$

$$\text{i.e., } n_{i.}\alpha_i - \sum_j n_{ij}\frac{\sum_l n_{lj}\alpha_l}{n_{.j}} = y_{i..} - \sum_j n_{ij}\frac{y_{.j.}}{n_{.j}}.$$

$$\text{i.e., } n_{i.}\alpha_i - \sum_l \alpha_l \frac{\sum_j n_{lj}n_{lj}}{n_{.j}} = \theta_i,$$

where $\theta_i = y_{i..} - \sum_j n_{ij}\frac{y_{.j.}}{n_{.j}}$. It is easily seen that $\sum_i \theta_i = 0$. Therefore imposing a suitable condition, we can solve for $\alpha_i$. Let $\widehat{\alpha}_i$ denote the solution. Then an estimate of $\beta_j$ is

$$\widehat{\beta}_j = \frac{y_{.j.}}{n_{.j}} - \frac{\sum_l n_{lj}\widehat{\alpha}_l}{n_{.j}} = \bar{y}_{.j.} - \frac{\sum_l n_{lj}\widehat{\alpha}_l}{n_{.j}}.$$

So the conditional minimum is

$$S_0^2 = \sum_{i,j,k} (y_{ijk} - \widehat{\alpha}_i - \widehat{\beta}_j)y_{ijk}$$

$$= \sum_{i,j,k}\left(y_{ijk} - \widehat{\alpha}_i - \frac{y_{.j.}}{n_{.j}} + \frac{\sum_l n_{lj}\widehat{\alpha}_l}{n_{.j}}\right) y_{ijk}$$

$$= \sum_{i,j,k} y_{ijk}^2 - \sum_{i,j,k}\widehat{\alpha}_i y_{ijk} - \sum_{i,j,k}\frac{y_{.j.}}{n_{.j}} y_{ijk} + \sum_{i,j,k}\frac{\sum_l n_{lj}\widehat{\alpha}_l}{n_{.j}} y_{ijk}$$

$$= \sum_{i,j,k} y_{ijk}^2 - \sum_{i}\widehat{\alpha}_i y_{i..} - \sum_{j}\frac{y_{.j.}^2}{n_{.j}} + \sum_{j}\frac{\sum_l n_{lj}\widehat{\alpha}_l}{n_{.j}} y_{.j.}$$

$$= \sum_{i,j,k} y_{ijk}^2 - \sum_{i}\widehat{\alpha}_i y_{i..} - \sum_{j}\frac{y_{.j.}^2}{n_{.j}} + \sum_{j}\sum_{l}\frac{n_{lj}\widehat{\alpha}_l}{n_{.j}} y_{.j.}$$

$$= \sum_{i,j,k} y_{ijk}^2 - \sum_{i}\widehat{\alpha}_i y_{i..} - \sum_{j}\frac{y_{.j.}^2}{n_{.j}} + \sum_{j}\sum_{i}\frac{n_{ij}\widehat{\alpha}_i}{n_{.j}} y_{.j.}$$

$$= \sum_{i,j,k} y_{ijk}^2 - \sum_{i}\widehat{\alpha}_i y_{i..} - \sum_{j}\frac{y_{.j.}^2}{n_{.j}} + \sum_{i}\widehat{\alpha}_i\sum_{j}\frac{n_{ij}}{n_{.j}} y_{.j.}$$

$$= \sum_{i,j,k} y_{ijk}^2 - \sum_{i}\widehat{\alpha}_i Q_i - \sum_{j}\frac{y_{.j.}^2}{n_{.j}}.$$

Therefore

$$S_0^2 - S^2 = \sum_{i,j}\frac{y_{ij.}^2}{n_{ij}} - \sum_{i}\widehat{\alpha}_i Q_i - \sum_{j}\frac{y_{.j.}^2}{n_{.j}}.$$

This is the S.S. due to interaction obtained by eliminating the block effects. The d.f. of $S_0^2$ is $n_{..} - pq + (p-1)(q-1) = n_{..} - p - q + 1$. So the d.f. of $S_0^2 - S^2$ is $n_{..} - p - q + 1 - n_{..} + pq = (p-1)(q-1)$. Under normality assumption the F-ratio to test for interaction is

$$F\big((p-1)(q-1), n_{..} - pq\big) = \frac{(S_0^2 - S^2)/(p-1)(q-1)}{S^2/(n_{..} - pq)}.$$

When F is not significant, we proceed to test the equality of the block and treatment effects. Consider $H_1: \alpha_1 = \alpha_2 = \cdots = \alpha_p$ $(= \alpha, \text{say.})$, a linear hypothesis of rank $(p-1)$. The unconditional minimum is $S_0^2$ with $(n_{..} - p - q + 1)$ d.f.. To obtain the conditional minimum we minimize

$$L_1 = \sum_{i,j,k} (y_{ijk} - \alpha - \beta_j)^2 \quad \text{w.r.t. } \alpha \text{ and } \beta_j\text{'s}$$

The normal equations are

$$\sum_{i,j,k} (y_{ijk} - \alpha - \beta_j) = 0$$

$$\text{and} \sum_{i,k} (y_{ijk} - \alpha - \beta_j) = 0, \;\; j = 1, \ldots, q.$$

$$\text{i.e., } y_{...} - n_{..}\alpha - \sum_j n_{.j.}\beta_j = 0 \tag{3}$$

$$\text{and } y_{.j.} - n_{.j}\alpha - n_{.j}\beta_j = 0 \tag{4}$$

From (4), $\widehat{\alpha + \beta_j} = \bar{y}_{.j.}$. Therefore the conditional minimum under $H_1$ is

$$S_0^2 = \sum_{i,j,k} \left( y_{ijk} - \frac{y_{.j.}}{n_{.j}} \right) y_{ijk}$$

$$= \sum_{i,j,k} y_{ijk}^2 - \sum_j \frac{y_{.j.}^2}{n_{.j}}$$

with $n_{..} - p - q + 1 + p - 1 = (n_{..} - q)$ d.f.. Therefore S.S. due to blocks after eliminating treatment effects is

$$S_1^2 - S_0^2 = \sum_i \hat{\alpha}_i Q_i$$

with d.f. $(p - 1)$. Similarly, if we consider the hypothesis

$$H_2\colon \beta_1 = \beta_2 = \cdots = \beta_q \; (= \beta, \text{say.})$$

the S.S. due to treatments can be found to be

$$S_2^2 - S_0^2 = \sum_j \frac{y_{.j.}^2}{n_{.j}} + \sum_i \hat{\alpha}_i Q_i - \sum_i \frac{y_{i..}^2}{n_{i.}} \quad \text{which has } (q-1) \text{ d.f.}$$

(The analysis of variance table is given below)

We first test the ratio $V_{T\times B}/V_E$. If it is significant we cannot proceed further. If it is not significant we can test the block effects by testing $V_B$ against $V_{T\times B}$ or $V_E$, whichever is greater. If $V_{T\times B}$ and $V_E$ are equal we take their weighted average $V_E'$ and test the F-ratio $V_B/V_E'$.

**Analysis of Variance Table**

| Variation due to | S.S. | d.f. | M.S.S. | F-ratio |
|---|---|---|---|---|
| Block (B)after eliminating treatments | $\sum \hat{\alpha}_i Q_i$ | $p-1$ | $V_B$ | |
| Treatments (T) ignoring blocks | $\sum_j \frac{y_{.j.}^2}{n_{.j}} - \frac{y_{...}^2}{n_{..}}$ | $q-1$ | | |
| Interaction (T × B) | $B' - T - B$ | $(p-1)(q-1)$ | $V_{T\times B}$ | $\frac{V_{T\times B}}{V_E}$ |
| Between cell (B′) | $\sum_{i,j} \frac{y_{ij.}^2}{n_{ij}} - \frac{y_{...}^2}{n_{..}}$ | $pq-1$ | | |
| Within cell or residual (R) | $T' - B'$ | $n_{..} - pq$ | $V_E$ | |
| Total (T′) | $\sum_{i,j,k} y_{ijk}^2 - \frac{y_{...}^2}{n_{..}}$ | $n_{..} - 1$ | | |
| Blocks ignoring treatments | $\sum_i \frac{y_{i..}^2}{n_{i.}} - \frac{y_{...}^2}{n_{..}}$ | $p-1$ | | |
| Treatments (P) eliminating blocks | $B' - T \times B - C$ | $q-1$ | $V_P$ | |

This table cannot be used to test the treatment effects because the S.S. due to treatments after ignoring blocks will contain block effects also. We have to add one more row in the table giving S.S. due to treatments (after eliminating blocks), which is given by

Between cell S. S. −S. S. due to interaction −
S. S. due to blocks ignoring treamtnes.

$$\text{i.e.}, P = B' - T \times B - C.$$

Then $V_P/V_E'$ will test for treatments.

If the interaction is significant we test the block effects for a fixed treatment and vice versa. In this case the procedure is the same as in a one way classification.

**Computational Procedure**

From the given data we form the following table:

| Treatments / Blocks | $B_1$ | $B_2$ | ... | $B_q$ | Block Total |
|---|---|---|---|---|---|
| $A_1$ | $y_{11.}$ | $y_{12.}$ | ... | $y_{1q.}$ | $y_{1..}$ |
| $A_2$ | $y_{21.}$ | $y_{22.}$ | ... | $y_{2q.}$ | $y_{2..}$ |
| ... | .. | ... | ... | ... | ... |
| $A_p$ | $y_{p1.}$ | $y_{p2.}$ | ... | $y_{pq.}$ | $y_{p..}$ |
| Treatment Total | $y_{.1.}$ | $y_{.2.}$ | ... | $y_{.q.}$ | $y_{...}$ |

We have to compute $\sum \hat{\alpha}_i Q_i$ where $Q_i = y_{i..} - \sum_j \frac{n_{ij}}{n_{.j}} y_{.j.}$. We have

$$n_{i.}\alpha_i - \sum_l \alpha_l \sum_j \frac{n_{ij}}{n_{.j}} n_{lj} = Q_i, \quad i = 1,2,\dots,p \tag{1}$$

Now, since $\sum \alpha_i = 0$, only $(p-1)$ of the equations are independent. Therefore we can impose a suitable condition to solve them, thus getting $\hat{\alpha}_1, \dots, \hat{\alpha}_p$.

Then equations (1) are called the adjusted equations, adjusted for the blocks effects.

## 3.12 Matrix Methods (when there is no interaction)

Let

$y = \left(y_{111}, y_{112}, \dots, y_{11n_{11}}, \dots, y_{pq1}, y_{pq2}, \dots, y_{pqn_{pq}}\right)$
and $\theta = (\alpha_1, \alpha_2, \dots, \alpha_p; \beta_1, \beta_2, \dots, \beta_q)$.

Then the observational equation is $E(y) = \theta A^T$ where

$A^T$

$$= \begin{bmatrix} J_{11} & J_{12} & \cdots & J_{1q} & o_{21} & o_{22} & \cdots & o_{2q} & \cdots & o_{p1} & o_{p2} & \cdots & o_{pq} \\ o_{11} & o_{12} & \cdots & o_{1q} & J_{21} & J_{22} & \cdots & J_{2q} & \cdots & o_{p1} & o_{p2} & \cdots & o_{pq} \\ \cdots & \cdots & \cdots & \cdots & \cdots & \cdots & \cdots & \cdots & \cdots & \cdots & \cdots & \cdots & \cdots \\ o_{11} & o_{12} & \cdots & o_{1q} & o_{21} & o_{22} & \cdots & o_{2q} & \cdots & J_{p1} & J_{p2} & \cdots & J_{pq} \\ J_{11} & o_{12} & \cdots & o_{1q} & J_{21} & o_{22} & \cdots & o_{2q} & \cdots & J_{p1} & o_{p2} & \cdots & o_{pq} \\ o_{11} & J_{12} & \cdots & o_{1q} & o_{21} & J_{22} & \cdots & o_{2q} & \cdots & o_{p1} & J_{p2} & \cdots & o_{pq} \\ \cdots & \cdots & \cdots & \cdots & \cdots & \cdots & \cdots & \cdots & \cdots & \cdots & \cdots & \cdots & \cdots \\ o_{11} & o_{12} & \cdots & J_{1q} & o_{21} & o_{22} & \cdots & J_{2q} & \cdots & o_{p1} & o_{p2} & \cdots & J_{pq} \end{bmatrix}$$

$$J_{ij} = (1, \dots, 1)_{1\times n_{ij}},$$

$$o_{ij} = (0, \dots, 0)_{1\times n_{ij}}, \quad \text{order of } A^T \text{ is } (p+q) \times \left(\sum\sum n_{ij}\right)$$

Sum of the first p rows = Sum of the last q rows.

So the rank of $A^T$ is only $(p+q-1)$ and hence there are only $(p+q-1)$ independent linear estimable parametric functions. A parametric function of the form

$$c\theta^T = c_1\alpha_1 + \cdots + c_p\alpha_p + c_1'\beta_1 + \cdots + c_q'\beta_q$$

is estimable if and only if $c_1 + \cdots + c_p = c_1' + \cdots + c_q'$. In particular contrasts are estimable.

From linear estimation theory we know that the least square normal equations giving $\theta$ are

$$yA = \theta A^T A,$$

and it can be seen that

$$AA^T = \begin{bmatrix} K & N \\ N^T & R \end{bmatrix}$$

where $K = \text{diag.}(n_{1.}, n_{2.}, \dots, n_{p.})$, $R = \text{diag.}(n_{.1}, n_{.2}, \dots, n_{.q})$ and $N = (n_{ij})$. The matrix N is called the incidence matrix.

Let $\theta = (\alpha|\beta)$ where $\alpha = (\alpha_1, \dots, \alpha_p)$, $\beta = (\beta_1, \dots, \beta_q)$, and

$$yA = \left(y_{1..}, y_{2..}, \dots, y_{p..} \middle| y_{.1.}, y_{.2.}, \dots, y_{.q.}\right)$$

$$= (B|T)$$

where

$$B = (B_1, B_2, \ldots, B_p) = (y_{1..}, y_{2..}, \ldots, y_{p..})$$

and

$$T = (T_1, T_2, \ldots, T_q) = (y_{.1.}, y_{.2.}, \ldots, y_{.q.}).$$

Then the normal equations become

$$(B|T) = (\alpha|\beta)\begin{bmatrix} K & N \\ N^T & R \end{bmatrix}$$

and since the partitioning is conformable for multiplication, we have

$$\alpha K + \beta N^T = B\,, \tag{1}$$

$$\alpha N + \beta R = T\,. \tag{2}$$

Note that K is non-singular since no $n_{i.} = 0$. Hence $K^{-1}$ exists. Post-multiplying (1) by $K^{-1}$, we get

$$\alpha + \beta N^T K^{-1} = BK^{-1}$$

$$\text{or } \alpha = BK^{-1} - \beta N^T K^{-1}.$$

Substituting in (2),

$$BK^{-1}N - \beta N^T K^{-1} N + \beta R = T$$

$$\text{i.e.,}\ \ \beta(R - N^T K^{-1} N) = T - BK^{-1}N \tag{3}$$

Now, $K^{-1}N = \left(\frac{n_{ij}}{n_{i.}}\right)$. So $N^T K^{-1} N = (\lambda_{ij})$ where $\lambda_{ij} = \sum_{k=1}^{p} n_{ki}\frac{n_{kj}}{n_{k.}}$.

Therefore

$$R - N^T K^{-1} N = \begin{bmatrix} n_{.1} - \lambda_{11} & -\lambda_{12} & \cdots & -\lambda_{1q} \\ -\lambda_{21} & n_{.2} - \lambda_{22} & \cdots & -\lambda_{2q} \\ \cdots & \cdots & \cdots & \cdots \\ -\lambda_{q1} & -\lambda_{q2} & \cdots & n_{.q} - \lambda_{qq} \end{bmatrix}$$

The sum of all the rows in $R - N^T K^{-1} N$ is zero and so it is singular. Since the rows satisfy one condition, the rank is only $(q - 1)$ and so we can impose one consistent condition and solve. Imposing a suitable condition, let $\hat{\beta}$ be a solution of (3) so that

$$\hat{\alpha} = BK^{-1} - \hat{\beta} N^T K^{-1}.$$

The unconditional minimum is

$$\begin{aligned} S^2 &= (y - \hat{\theta}A^T)y^T \\ &= yy^T - \hat{\theta}A^Ty^T \\ &= yy^T - (\hat{\alpha}|\hat{\beta})\begin{bmatrix} B^T \\ T^T \end{bmatrix} \\ &= yy^T - \hat{\alpha}B^T - \hat{\beta}T^T \\ &= yy^T - (BK^{-1} - \hat{\beta}N^TK^{-1})B^T - \hat{\beta}T^T \\ &= yy^T - BK^{-1}B^T + \hat{\beta}N^TK^{-1}B^T - \hat{\beta}T^T \\ &= yy^T - BK^{-1}B^T - \hat{\beta}(T^T - N^TK^{-1}B^T) \\ &= yy^T - BK^{-1}B^T - \hat{\beta}Q^T \end{aligned}$$

where $Q = T - BK^{-1}N$. The total number of observations is $n_{..}$ and the rank of A is $(p + q - 1)$. So the d. f. of $S^2$ is $n_{..} - (p + q - 1)$.

Now, consider the hypothesis

$$H_0: \beta_1 = \beta_2 = \cdots = \beta_q, (= \beta_0, \text{say})$$

a linear hypothesis of rank $(q - 1)$. Now

$$\theta = (\alpha_1, \ldots, \alpha_p, \beta_0).$$

The observational equations are

$$E(y) = \theta A_0^T,$$

where

$$A_0^T = \begin{bmatrix} J_1 & o_2 & \cdots & o_p \\ o_1 & J_2 & \cdots & o_p \\ \cdots & \cdots & \cdots & \cdots \\ o_1 & o_2 & \cdots & J_p \\ J_1 & J_2 & \cdots & J_p \end{bmatrix}$$

where

$$J_i = (1,1, \ldots, 1)_{1\times n_i}, \quad o_i = (0,0, \ldots, 0)_{1\times n_i}.$$

$$A_0^T A_0 = \begin{pmatrix} n_{1.} & 0 & \cdots & 0 & n_{1.} \\ 0 & n_{2.} & \cdots & 0 & n_{2.} \\ \cdots & \cdots & \cdots & \cdots & \cdots \\ 0 & 0 & \cdots & n_{p.} & n_{p.} \\ n_{1.} & n_{2.} & \cdots & n_{p.} & n_{..} \end{pmatrix}$$

and Rank $A_0 = p$. Therefore

$$\theta A_0^T A_0 = [(\alpha_1 + \beta_0)n_{1.}, (\alpha_2 + \beta_0)n_{2.}, \ldots, (\alpha_p + \beta_0)n_{p.}, \sum \alpha_r n_{r.} + n_{..}\beta_0]$$

$$yA_0 = (y_{1..}, y_{2..}, \ldots, y_{p..}, y_{...}) = (B_1, B_2, \ldots, B_p, y_{...}).$$

The least square normal equations are $yA_0 = \theta A_0^T A_0$, i.e.,

$$(B_1, B_2, \ldots, B_p, y_{...}) = \left[(\alpha_1 + \beta_0)n_{1.}, (\alpha_2 + \beta_0)n_{2.}, \ldots, (\alpha_p + \beta_0)n_{p.}, \sum \alpha_r n_{r.} + n_{..}\beta_0\right].$$

Out of the $(p + 1)$ equations here only p are independent since rank of $A_0 = p$. So we can impose one condition and solve. Let us impose the condition $\beta_. = 0$, which is equivalent to omitting the last row and last column in $A_0^T A_0$. The equations become

$$\left(\alpha_1 n_{1.}, \alpha_2 n_{2.}, \ldots, \alpha_p n_{p.}, \sum \alpha_r n_{r.}\right) = (B_1\ B_2 \ldots B_p\ y_{...}).$$

The first p equations are given by

$$\alpha K = B$$

giving $\hat{\alpha} = BK^{-1}$.

Thus a solution for $\theta$ is $\hat{\theta} = (\hat{\alpha}\ \ 0)$. So the conditional minimum S.S. is

$$\begin{aligned} S_0^2 &= (y - \hat{\theta}A_0^T)y^T \\ &= yy^T - \hat{\theta}A_0^T y^T \\ &= yy^T - (\hat{\alpha}\ \ 0)\begin{pmatrix} B^T \\ y_{...} \end{pmatrix} \\ &= yy^T - \hat{\alpha}B^T \end{aligned}$$

$$= yy^T - BK^{-1}B^T,$$

which has got $n_{..} - (p + q - 1) + q - 1 = (n_{..} - p)$ d.f.. Therefore the S.S. due to treatments eliminating block effects is

$$S_0^2 - S^2 = \hat{\beta}\theta^T$$

which has $(q - 1)$ d.f..

**Analysis of variance table**

| Variation Due to | S.S. | d.f. | M.S.S. | F |
|---|---|---|---|---|
| Blocks (ignoring treatments) | $BK^{-1}B^T - \frac{y_{...}^2}{n_{..}}$ | $p - 1$ | | |
| Treatments (eliminating blocks) | $\hat{\beta}\theta^T$ | $q - 1$ | $V_T$ | $V_T/V_E$ |
| Residual | (By sub) | (Sub) | $V_E$ | |
| Total | $yy^T - \frac{y_{...}^2}{n_{..}}$ | $n_{..} - 1$ | | |

The table provides a test for the treatment effects when there is no interaction.

Note:

The particular cases of the above case are

(i) When each $n_{ij}$ is unity

(ii) When the $n_{ij}$'s are all equal, say $n_{ij} = n$

(iii) When the $n_{ij}$'s are proportional.

### 3.13 Case of Proportional Frequencies

The procedure can be simplified to a great extent if the $n_{ij}$'s are proportional, i.e., if

$$\frac{n_{ij}}{n_{lj}} = \frac{n_{i.}}{n_{l.}}, \quad j = 1, \ldots, q$$

$$\text{i.e.,} \quad n_{ij}n_{l.} = n_{i.}n_{lj}\,.$$

Summing w.r.t. l,

$$n_{ij} n_{..} = n_{i.} n_{.j} \quad \text{so that} \quad n_{ij} = \frac{n_{i.} n_{.j}}{n_{..}}.$$

Let us find $\hat{\alpha}_i$ in this case. We have

$$\sum_l \alpha_l \sum_j \frac{n_{ij} n_{lj}}{n_{.j}} = \sum_l \alpha_l \sum_j \frac{n_{i.} n_{.j}}{n_{..}} \times \frac{n_{l.} n_{.j}}{n_{..}} \times \frac{1}{n_{.j}}$$

$$= \sum_l \alpha_l \frac{n_{i.} n_{l.}}{n_{..}^2} \times \sum_j n_{.j}$$

$$= \sum_l \alpha_l \frac{n_{i.} n_{l.}}{n_{..}}$$

$$= \frac{n_{i.}}{n_{..}} \sum_l \alpha_l \, n_{l.}\,.$$

Again,

$$Q_i = y_{i..} - \sum_j \frac{n_{ij}}{n_{.j}} y_{.j.}$$

$$= y_{i..} - \frac{n_{i.}}{n_{..}} y_{...}.$$

Therefore the equations giving $\alpha_i$'s become

$$n_{i.} \alpha_i - \frac{n_{i.}}{n_{..}} \sum_l \alpha_l \, n_{l.} = y_{i..} - \frac{n_{i.}}{n_{..}} y_{...}$$

$$\text{i.e.,} \quad \alpha_i - \frac{1}{n_{..}} \sum_l \alpha_l \, n_{l.} = \bar{y}_{i..} - \bar{y}_{...}; \quad i = 1, \ldots, p\,.$$

Since only $(p-1)$ of these equations are independent we can impose one suitable condition and solve. Let the condition imposed be $\sum_l \alpha_l \, n_{l.} = 0$. Then

$$\hat{\alpha}_i = \bar{y}_{i..} - \bar{y}_{...},$$

which is the same as in the case when the observations in the cells are the same.

**Case when there is interaction**

In this case there is no exact test for the blocks and treatments. But we can test the block effects for fixed treatment effects and vice-versa. This process will help to reduce the effect due to interaction.

Let us consider the treatment $T_j$. The observations will be as in the following table (with n observations per cell.):

| Cell | Treatment $T_j$ | Sum |
|---|---|---|
| 1 | $y_{1j1}$<br>$y_{1j2}$<br>...<br>$y_{1jn}$ | $y_{1j.}$ |
| 2 | $y_{2j1}$<br>$y_{2j2}$<br>...<br>$y_{2jn}$ | $y_{2j.}$ |
| ... | ... | ... |
| p | $y_{pj1}$<br>$y_{pj2}$<br>...<br>$y_{pjn}$ | $y_{pj.}$ |

**Analysis of variance table:**

| Variation Due to | S.S. | d.f. | M.S.S. | F |
|---|---|---|---|---|
| Blocks (for fixed treatment $T_j$) | $\sum_{i,j} \frac{y_{ij.}^2}{n} - \sum_{i} \frac{y_{ij.}^2}{n}$ | $p-1$ | $V_B$ | $V_B/V_E$ |
| Between cell | $\sum_{i,j} \frac{y_{ij.}^2}{n} - \frac{y_{...}^2}{npq}$ | | | |
| Within cell (Residual) | (By sub) | (Sub) | $V_E$ | |

| Total | $\sum y_{ijk}^2 - \frac{y_{...}^2}{npq}$ | $npq - 1$ | | |
|---|---|---|---|---|

$F = V_B/V_E$ will test the block effects for the j-th treatment.

**3.14 Three-Way Classification**

Let there be pqr observations classified according to one set of p characteristics $A_1, \dots, A_p$, a second set of q characteristics $B_1, \dots, B_q$ and a third set of r characteristics $C_1, \dots, C_r$. For each characteristic of one set we get a two-way classification of the other two characteristics. Let us consider the simplest case of a single observation per cell. Let $y_{ijk}$ be the observation in the (i,j)-th cell of the two-way classification corresponding to the characteristic $C_k$.

Let $\alpha_i$ be the effect of $A_i$, $\beta_j$ be the effect of $B_j$, $\gamma_k$ be the effect of $C_k$, $a_{ij}$ be the effect of the interaction due to the $(AB) \times C$ classification, $b_{jk}$ be the effect of the interaction due to the $(BC) \times A$ classification, and $c_{ki}$ be the effect of the interaction due to the $(CA) \times B$ classification.

[If the A-B classification changes for a fixed C-classification, then it is due to the interaction between the A and B classifications. It is this interaction that we denoted by $a_{ij}$. So $a_{ij}$ is the combined effect of the two characteristics $A_i$ and $B_j$.]

Assuming additivity of effects, we have

$$y_{ijk} = \alpha_i + \beta_j + \gamma_k + a_{ij} + b_{jk} + c_{ki} + \varepsilon_{ijk},$$

where $\varepsilon_{ijk}$ is a random error which we assume to be normally distributed with mean zero and variance $\sigma^2$. The total error S.S. is

$$L = \sum_{i,j,k} (y_{ijk} - \alpha_i - \beta_j - \gamma_k - a_{ij} - b_{jk} - c_{ki})^2$$

Differentiating L w.r.t. each of the parameters and equating to zero we get the least square normal equations to be (after necessary simplifications),

$$y_{i..} - qr\alpha_i - r\beta_. - q\gamma_. - ra_{i.} - b_{..} - qc_{.i} = 0 \quad (1)$$

$$y_{.j.} - r\alpha_. - pr\beta_j - p\gamma_. - ra_{.j} - pb_{j.} - c_{..} = 0 \quad (2)$$

$$y_{..k} - q\alpha_. - p\beta_. - pq\gamma_k - a_{..} - pb_{.k} - qc_{k.} = 0 \quad (3)$$

$$y_{ij.} - r\alpha_i - r\beta_j - \gamma_. - ra_{ij} - b_{j.} - c_{.i} = 0 \quad (4)$$

$$y_{.jk} - \alpha_. - p\beta_j - p\gamma_k - a_{.j} - pb_{jk} - c_{k.} = 0 \quad (5)$$

$$y_{i.k} - q\alpha_i - \beta_. - q\gamma_k - a_{i.} - b_{.k} - qc_{ki} = 0 \quad (6)$$

$$i = 1,2,\ldots,p, \;\; j = 1,2,\ldots,q, \;\; k = 1,2,\ldots,r.$$

Altogether there are $p+q+r+pq+qr+pr$ equations. Let us get the total number of independent equations.

$$\sum_j (4) = (1), \;\; \sum_i (4) = (2) \;\text{ and }\; \sum_j (5) = (3).$$

Thus the first three sets of equations depend on the last three sets and so we may discard them. There remain $pq+qr+pr$ equations. Again,

$$\sum_i (4) = \sum_k (5), \quad \text{for any j},$$

$$\sum_j (5) = \sum_i (6), \quad \text{for any k},$$

$$\text{and} \quad \sum_k (6) = \sum_j (4), \quad \text{for any i}.$$

Further, $\Sigma_{i,j}(4) = \Sigma_{j,k}(5) = \Sigma_{i,k}(6)$. There are $(p+r+q-1)$ conditions and so the total number of independent normal equations is

$$(pq+qr+pr) - (p+q+r-1).$$

Altogether the normal equations satisfy $2(p+q+r)-1$ conditions. So we can impose $2(p+q+r)-1$ conditions to solve them. Let us impose the conditions

$$\sum_i a_{ij} = \sum_j a_{ij} = 0 \quad (p + q - 1 \text{ conditions})$$

$$\sum_j b_{jk} = \sum_k b_{kj} = 0 \quad (q + r - 1 \text{ conditions})$$

$$\sum_k c_{ki} = \sum_i c_{ki} = 0 \quad (p + r - 1 \text{ conditions})$$

and $\beta_. = \gamma_. = 0$ (2 conditions). These conditions imply $a_{..} = 0, b_{..} = 0, c_{..} = 0$. The estimates are now given by

$$\hat{\alpha}_i = \frac{y_{i..}}{qr}, \text{ from (1)},$$

$$\hat{\beta}_j = \frac{y_{.j.}}{pr} - \frac{y_{...}}{pqr}, \text{ from (2), since } \alpha_. = \sum_i \alpha_i = \frac{y_{...}}{qr}$$

$$\hat{\gamma}_k = \frac{y_{..k}}{pq} - \frac{y_{...}}{pqr}, \text{ from (3)},$$

$$\hat{a}_{ij} = \frac{y_{ij.}}{r} - \frac{y_{i..}}{qr} - \frac{y_{.j.}}{pr} + \frac{y_{...}}{pqr}, \text{ from (4)}$$

$$\hat{b}_{jk} = \frac{y_{.jk}}{p} - \frac{y_{.j.}}{pr} - \frac{y_{..k}}{pq} + \frac{y_{...}}{pqr}, \text{ from (5)}$$

$$\hat{c}_{ki} = \frac{y_{i.k}}{q} - \frac{y_{i..}}{qr} - \frac{y_{..k}}{pq} + \frac{y_{...}}{pqr}, \text{ from (6)}.$$

Therefore the unconditional minimum S.S. is, after simplification,

$$S^2 = \sum_{i,j,k} (y_{ijk} + \frac{y_{i..}}{qr} + \frac{y_{.j.}}{pr} + \frac{y_{..k}}{pq} - \frac{y_{ij.}}{r} - \frac{y_{.jk}}{p} - \frac{y_{i.k}}{q} - \frac{y_{...}}{pqr})^2$$

and it has $pqr - (pq + qr + pr) + (p + q + r - 1) = (p - 1)(q - 1)(r - 1)$ d.f..

**(i) Consider the hypothesis $H_0$:** $a_{ij} = 0$

This is a linear hypothesis of rank $(p-1)(q-1)$. The total error S.S. under $H_0$ is

$$L_0 = \sum_{i,j,k} (y_{ijk} - \alpha_i - \beta_j - \gamma_k - b_{jk} - c_{ki})^2.$$

The normal equations are

$$y_{i..} - qr\alpha_i - r\beta_. - q\gamma_. - b_{..} - qc_{.i} = 0 \qquad (7$$

$$y_{.j.} - r\alpha_. - pr\beta_j - p\gamma_. - pb_{j.} - c_{..} = 0 \qquad (8)$$

$$y_{..k} - q\alpha_. - p\beta_. - pq\gamma_k - pb_{.k} - qc_{k.} = 0 \qquad (9)$$

$$y_{.jk} - \alpha_. - p\beta_j - p\gamma_k - pb_{jk} - c_{k.} = 0 \qquad (10)$$

$$y_{i.k} - q\alpha_i - \beta_. - q\gamma_k - b_{.k} - qc_{ki} = 0 \qquad (11)$$

There are $(p+q+r+qr+pr)$ equations. But since $\sum_k(11) = (7), \sum_k(10) = (8), \sum_j(10) = (9)$, we discard the first $p+q+r$ equations leaving $(qr+pr)$ equations. Again, $\sum_j(iv) = \sum_i(v)$. This involves r conditions. Hence the total number of independent equations is $pr + qr - r = (p+q-1)r$. Since the normal equations satisfy $p+q+2r$ independent conditions we may introduce $p+q+2r$ conditions to solve them. Let them be

$$\sum_j b_{jk} = 0 = \sum_k b_{jk} \quad (q+r-1 \text{ conditions})$$

$$\sum_i c_{ki} = 0 = \sum_k c_{ki} \quad (p+r-1 \text{ conditions})$$

and $\beta_. = \gamma_. = 0$ (2 conditions). These imply $b_{..} = c_{..} = 0$. The least square estimates are now given by

$$\widehat{\alpha}_i = \frac{y_{i..}}{qr}, \quad \text{from (7)},$$

$$\widehat{\beta}_j = \frac{y_{.j.}}{pr} - \frac{y_{...}}{pqr}, \quad \text{from (8)},$$

$$\widehat{\gamma}_k = \frac{y_{..k}}{pq} - \frac{y_{...}}{pqr}, \quad \text{from (9)},$$

$$\widehat{b}_{jk} = \frac{y_{.jk}}{p} - \frac{y_{.j.}}{pr} - \frac{y_{..k}}{pq} + \frac{y_{...}}{pqr}, \quad \text{from (10)}$$

$$\widehat{c}_{ki} = \frac{y_{i.k}}{q} - \frac{y_{i..}}{qr} - \frac{y_{..k}}{pq} + \frac{y_{...}}{pqr}, \quad \text{from (11)}.$$

The conditional minimum S.S. then is

$$S_0^2 = \sum_{i,j,k}\left(y_{ijk} - \frac{y_{.jk}}{p} - \frac{y_{i.k}}{q} + \frac{y_{...}}{pqr}\right)^2$$

$$= \sum_{i,j,k}\left(y_{ijk} + \frac{y_{i..}}{qr} + \frac{y_{.j.}}{pr} + \frac{y_{..k}}{pq} - \frac{y_{ij.}}{r} - \frac{y_{i.k}}{q} - \frac{y_{.jk}}{p} - \frac{y_{...}}{pqr}\right)^2$$

$$+ \sum_{i,j,k}\left(\frac{y_{i..}}{qr} + \frac{y_{.j.}}{pr} - \frac{y_{ij.}}{r} - \frac{y_{...}}{pqr}\right)^2.$$

Therefore the S.S. due to the $A \times B$ interaction is

$$S_0^2 - S^2 = \sum_{i,j,k}\left(\frac{y_{i..}}{qr} + \frac{y_{.j.}}{pr} - \frac{y_{ij.}}{r} - \frac{y_{...}}{pqr}\right)^2$$

$$= r\sum_{i,j}\left(\frac{y_{i..}}{qr} + \frac{y_{.j.}}{pr} - \frac{y_{ij.}}{r} - \frac{y_{...}}{pqr}\right)^2$$

with $(p-1)(q-1)$ d.f.. Therefore under normality assumptions, the F-ratio to test the $A \times B$ interaction is

$$F\big((p-1)(q-1), (p-1)(q-1)(r-1)\big)$$

$$= \frac{(S_0^2 - S^2)/(p-1)(q-1)}{S^2/(p-1)(q-1)(r-1)}.$$

Similarly, the F-ratio to test the $B \times C$ and $C \times A$ interactions are, respectively,

$$F\big((q-1)(r-1),(p-1)(q-1)(r-1)\big) = \frac{(S_1^2 - S^2)/(q-1)(r-1)}{S^2/(p-1)(q-1)(r-1)}$$

and

$$F\big((p-1)(r-1),(p-1)(q-1)(r-1)\big) = \frac{(S_2^2 - S^2)/(p-1)(r-1)}{S^2/(p-1)(q-1)(r-1)}$$

where

$$S_1^2 - S^2 = p\sum_{j,k}\left(\frac{y_{.j.}}{pr} + \frac{y_{..k}}{pq} - \frac{y_{.jk}}{p} - \frac{y_{...}}{pqr}\right)^2$$

and

$$S_2^2 - S^2 = \sum_{i,k}\left(\frac{y_{i..}}{qr} + \frac{y_{..k}}{pq} - \frac{y_{i.k}}{q} - \frac{y_{...}}{pqr}\right)^2.$$

**(ii). Consider the hypothesis $H_0'$: $\alpha_1 = \cdots = \alpha_p$ (= $\alpha$, say)**

The total error S.S. is

$$L_0' = \sum_{i,j,k}(y_{ijk} - \alpha - \beta_j - \gamma_k - a_{ij} - b_{jk} - c_{ki})^2.$$

Proceeding as before, let ${S_0'}^2$ be the minimum error S.S.. Then the S.S. due to the A-effects will be

$${S_0'}^2 - S^2 = qr\sum_i\left(\frac{y_{i..}}{qr} - \frac{y_{...}}{pqr}\right)^2$$

with $(p-1)$ d.f. so that the test criterion for the A-effects will be

$$F\big((p-1),(p-1)(q-1)(r-1)\big) = \frac{({S_0'}^2 - S^2)/(p-1)}{S^2/(p-1)(q-1)(r-1)}$$

We get similar formula for the B and C-effects above.

**Analysis of Variance Table**

| Variation Due to | S.S. | d.f. | M.S.S | F |
|---|---|---|---|---|
| A effects | $qr\sum_i(\frac{y_{i..}}{qr}-\frac{y_{...}}{pqr})^2$ | $p-1$ | $V_A$ | $\frac{V_A}{V_E}=F_1$ |
| B effects | $pr\sum_j(\frac{y_{.j.}}{pr}-\frac{y_{...}}{pqr})^2$ | $q-1$ | $V_B$ | $\frac{V_B}{V_E}=F_2$ |
| C effects | $pq\sum_k(\frac{y_{..k}}{pq}-\frac{y_{...}}{pqr})^2$ | $r-1$ | $V_C$ | $\frac{V_C}{V_E}=F_3$ |
| A × B interaction | $r\sum_{i,j}(\frac{y_{i..}}{qr}+\frac{y_{.j.}}{pr}-\frac{y_{ij.}}{r}-\frac{y_{...}}{pqr})^2$ | $(p-1)(q-1)$ | $V_{AB}$ | $\frac{V_{AB}}{V_E}=F_4$ |
| B × C interaction | $p\sum_{j,k}(\frac{y_{.j.}}{pr}+\frac{y_{..k}}{pq}-\frac{y_{.jk}}{p}-\frac{y_{...}}{pqr})^2$ | $(q-1)(r-1)$ | $V_{BC}$ | $\frac{V_{BC}}{V_E}=F_5$ |
| C × A interaction | $q\sum_{i,k}(\frac{y_{.j.}}{qr}+\frac{y_{..k}}{pq}-\frac{y_{i.k}}{q}-\frac{y_{...}}{pqr})^2$ | $(p-1)(r-1)$ | $V_{AC}$ | $\frac{V_{AC}}{V_E}=F_6$ |
| Error (also called A × B × C interaction) | By Subtraction | $(p-1)(q-1)(r-1)$ | $V_E$ | |
| Total | $\sum_{i,j,k}(y_{ijk}-\frac{y_{...}}{pqr})^2$ | $pqr-1$ | | |

Note: we can test the significance of $F_1, F_2, F_3$ since the S.S. due to the A, B, C effects are not affected by the $A \times B, B \times C$ or $A \times C$ interactions, also called first order interactions. $A \times B \times C$ is called a second order interaction.

## 3.15 Randomized Block Design (RBD)

This is the simplest design in analysis and set up. It is usually applicable when the number of treatments to be compared is small, say less than 15, and when it is possible to get uniform blocks which can be divided into sub-plots to apply all the treatments in a block.

The set up of the design is as follows:

We have v treatments $T_1, T_2, \dots, T_v$ and r blocks $B_1, B_2, \dots, B_r$. Each block is divided into v plots and the v treatments are applied into these sub-plots, the choice being dicided by a random process.

Let $y_{ij}$ be the yield of the j-th treatment in the i-th block. The yield $y_{ij}$ is the sum of three effects – a general effect g, effect of the i-th block $b_i$, and effect of the j-th treatment $t_j$. In symbols,

$$y_{ij} = g + b_i + t_j + \varepsilon_{ij}$$

where $\varepsilon_{ij}$ is an error which we assume to be distributed normally around zero with a common but unknown variance $\sigma^2$. We further assume that the blocks are uniform, i.e., between plots in the same block there are no fertility differences.

The observational equations are

$$E(y_{ij}) = g + b_i + t_j; \quad i = 1, \dots, r, \quad j = 1, \dots, v.$$

Therefore the total error S.S. is

$$L = \sum_{i,j} (y_{ij} - g - b_i - t_j)^2.$$

We estimate the parameters by minimizing L. The normal equations are (in usual notations),

$$\bar{y}_{..} = g + \frac{t_{.}}{v} + \frac{b_{..}}{r} \tag{1}$$

$$\bar{y}_{i.} = g + b_i + \frac{t_{.}}{v}, \quad i = 1, \dots, r \tag{2}$$

$$\bar{y}_{.j} = g + \frac{b_{..}}{r} + t_j, \quad j = 1, \dots, v \tag{3}$$

There are $r + v + 1$ equations. But since $\sum_i(2) = \sum_j(3) = (1)$, only $n + v - 1$ equations are independent. Thus we may introduce two conditions to solve them. Let them be $b_{.} = 0, t_{.} = 0$. Then the solutions are

$$\hat{g} = \bar{y}_{..}, \quad \hat{b}_i = \bar{y}_{i.} - \bar{y}_{..}, \quad \hat{t}_j = \bar{y}_{.j} - \bar{y}_{..}.$$

Therefore the minimum error S.S. is

$$S^2 = \sum_{i,j} (y_{ij} - \hat{g} - \hat{b}_i - \hat{t}_j)^2$$

$$= \sum_{i,j} (y_{ij} - \bar{y}_{i.} - \bar{y}_{.j} + \bar{y}_{..})^2$$

with $rv - r - v + 1 = (r-1)(v-1)$ d.f..

Suppose we want to test for the difference between treatment effects. The hypothesis is

$$H_0: t_1 = t_2 = \cdots = t_v \ (= t, \text{say})$$

which is of rank $v - 1$. The error S.S. under $H_0$ is

$$L_0 = \sum_{i,j} (y_{ij} - g - b_i - t)^2.$$

The least square normal equations are

$$\sum_{i,j} (y_{ij} - g - b_i - t) = 0$$

$$\sum_{j} (y_{ij} - g - b_i - t) = 0$$

$$\sum_{i,j} (y_{ij} - g - b_i - t) = 0$$

Considering the first two equations,

$$\bar{y}_{..} = g + t + \frac{b_.}{r} \tag{4}$$

$$\text{and } \bar{y}_{i.} = g + t + b_i \tag{5}$$

Since (4) and (5) satisfy one condition we may put $b_. = 0$ and solve. Then the estimates are

$$\widehat{(g+t)} = \bar{y}_{..} \ \text{ and } \ b_i = \bar{y}_{i.} - \bar{y}_{..}\,.$$

Therefore the conditional minimum S.S. is

$$S_0^2 = \sum_{i,j} (y_{ij} - \bar{y}_{i.} + \bar{y}_{..} - \bar{y}_{..})^2 = \sum_{i,j} (y_{ij} - \bar{y}_{i.})^2$$

which has $rv - r - v + 1 + v - 1 = r(v-1)$ d.f.. Therefore the S.S. due to treatments is $S_0^2 - S^2$ with $v - 1$ d·f·. Therefore the test criterion for treatment effects is

$$F\big(v-1,(r-1)(v-1)\big) = \frac{(S_0^2 - S^2)/(v-1)}{S^2/(r-1)(v-1)}.$$

Similarly the test criterion for block effect is

$$F\big(r-1,(r-1)(v-1)\big) = \frac{(S_1^2 - S^2)/(r-1)}{S^2/(r-1)(v-1)}$$

where $S_1^2 = \sum_{i,j}(y_{ij} - \bar{y}_{.j})^2$.

Now the total S.S. (with $rv - 1$ d.f.) is

$$\sum_{i,j} (y_{ij} - \bar{y}_{..})^2 = \sum_{i,j} [(y_{ij} - \bar{y}_{i.} - \bar{y}_{.j} + \bar{y}_{..}) + (\bar{y}_{i.} - \bar{y}_{..})$$

$$+ (\bar{y}_{.j} - \bar{y}_{..})]^2$$

$$= \sum_{i,j} (y_{ij} - \bar{y}_{i.} - \bar{y}_{.j} + \bar{y}_{..})^2 + \sum_{i,j} (\bar{y}_{i.} - \bar{y}_{..})^2$$

$$+ \sum_{i,j} (\bar{y}_{.j} - \bar{y}_{..})^2,$$

$$\text{(since the cross} - \text{product terms vanish.)}$$

$$= \sum_{i,j} (y_{ij} - \bar{y}_{i.} - \bar{y}_{.j} + \bar{y}_{..})^2 + v\sum_{i} (\bar{y}_{i.} - \bar{y}_{..})^2$$

$$+ r\sum_{j} (\bar{y}_{.j} - \bar{y}_{..})^2$$

$$= \text{Error S.S.} + \text{Block S.S.} + \text{Treatment S.S.}$$

**Analysis of Variance Table**

| Source | S.S. | d.f. | M.S.S. | F |
|---|---|---|---|---|
| Blocks | $v\sum_{i}(\bar{y}_{i.}-\bar{y}_{..})^2$ | $r-1$ | $V_B$ | $V_B/V_E$ |
| Treatments | $r\sum_{j}(\bar{y}_{.j}-\bar{y}_{..})^2$ | $v-1$ | $V_T$ | $V_T/V_E$ |
| Error | (Sub) | (Sub) | $V_E$ | |
| Total | $\sum_{i,j}(y_{ij}-\bar{y}_{..})^2$ | $rv-1$ | | |

The ratios $V_B/V_E$ and $V_T/V_E$ will test the differences in the block and treatment effects.

**Computations**

$$\sum_{i,j}(y_{ij}-\bar{y}_{..})^2 = \sum_{i,j} y_{ij}^2 - \bar{y}_{..}^2\,,$$

$$\sum_{i,j}(\bar{y}_{i.}-\bar{y}_{..})^2 = v\sum_{i}(\bar{y}_{i.}-\bar{y}_{..})^2$$

$$= v\sum_{i}\left(\frac{y_{i.}}{v}-\frac{y_{..}}{rv}\right)^2$$

$$= \sum_{i}\frac{y_{i.}^2}{v}-\frac{y_{..}^2}{rv},$$

and

$$\sum_{i,j}(\bar{y}_{.j}-\bar{y}_{..})^2 = \sum_{j}\frac{y_{.j}^2}{r}-\frac{y_{..}^2}{rv},$$

## 3.16 Judging the Relative Merits of Treatment

If the test criterion $V_T/V_E$ gives a significant result, then the treatment effects $t_1, \dots, t_v$ are significantly different. To judge which is the best treatment, we have to test the significance of differences of the type $t_j - t_k$. We have

$$\hat{t}_j - \hat{t}_k = \bar{y}_{.j} - \bar{y}_{.k}$$

where $\bar{y}_{.j}$ and $\bar{y}_{.k}$ are independent and each has variance $\frac{\sigma^2}{r}$. Therefore

$$V(\bar{y}_{.j} - \bar{y}_{.k}) = \frac{2\sigma^2}{r}.$$

Best estimate of $\sigma^2$ is the variance due to error, i.e., $V_E$. Under the hypothesis $t_j = t_k$,

$$\frac{\bar{y}_{.j} - \bar{y}_{.k}}{\sqrt{\frac{2V_E}{r}}}$$

is a Students' t with $(r-1)(v-1)$ d. f.. If $t_0$ is the 5% tabled value of t, then the treatment $t_j$ and $t_k$ are significantly different if

$$|\bar{y}_{.j} - \bar{y}_{.k}| \geq t_0 \sqrt{\frac{2V_E}{r}}.$$

## 3.17 Confidence Limits for Treatment Differences

Let $t_j - t_k = d$. Then

$$\frac{(\hat{t}_j - \hat{t}_k) - (t_j - t_k)}{\sqrt{\text{Estimate of } V(\hat{t}_j - \hat{t}_k)}}$$

is a Student's t with $(r-1)(v-1)$ d.f.. Estimate of $V(\hat{t}_j - \hat{t}_k) = \frac{2V_E}{r}$.

Let $t_0$ be the $\alpha$% tabled value of t for $(r-1)(v-1)$ d.f.. Then

$$(\bar{y}_{.j} - \bar{y}_{.k}) \pm t_0 \sqrt{\frac{2V_E}{r}}$$

are the $(1-\alpha)\%$ confidence limits for $t_j - t_k$.

## 3.18 Uniformity Trial (cf. R.A.Fisher and Cochran)

We have assumed that there is no heterogeneity within blocks. Therefore we have to get blocks with homogeneous plots. Thus is mostly achieved through experience. Where experience fails we adopt what is called uniformity trial. We divide the blocks into a number of plots and in each plot apply the same treatment. Let the yields in the plots be tabulated as follows:

$$\begin{matrix} y_{11} & y_{12} & \cdots & y_{1v} \\ y_{21} & y_{22} & \cdots & y_{2v} \\ \cdots & \cdots & \cdots & \cdots \\ y_{r1} & y_{r2} & \cdots & y_{rv} \end{matrix}$$

Assuming that there are row and column effects, we may analysis the experiment as follows:

| Source | S.S. | d.f. | M.S.S. | F |
|---|---|---|---|---|
| Rows | $\sum_i \frac{y_{i.}^2}{r} - \frac{y_{..}^2}{rv}$ | $r-1$ | $V_R$ | $F_1 = V_R/V_E$ |
| Columns | $\sum_j \frac{y_{.j}^2}{v} - \frac{y_{..}^2}{rv}$ | $v-1$ | $V_C$ | $F_2 = V_C/V_E$ |
| Residual | (Sub) | (Sub) | $V_E$ | |
| Total | $\sum_{i,j} y_{ij}^2 - \frac{y_{..}^2}{rv}$ | $rv-1$ | | |

Suppose $F_1$ is significant (i.e., between rows there is heterogeneity) and $F_2$ is not significant (i.e., between columns there is no heterogeneity.) Then we may take the rows as blocks and design the experiment.

If $F_1$ is not significant and $F_2$ is significant, then we may take the columns as blocks.

If both $F_1$ and $F_2$ are significant, then the randomized block experiment cannot be designed on these plots.

If $F_1$ and $F_2$ are both not significant, the experiment may be conducted on these plots and it will be found that the between block variation is insignificant.

The uniformity trial is both expensive and time-consuming. Therefore we resort to it only if experience fails in getting a set of homogenous blocks.

## 3.19 Latin Square Design

This is a design which achieves a two-way elimination of heterogeneity.

A plot of land nearly square in shape is divided into an equal number of rows and columns. the treatments are assigned to the plots in a random manner subject to the condition that each row and column receives the treatments once and only once. As an example of a 5-sided Latin square will be (with numbers indicating treatments)

| | | | | |
|---|---|---|---|---|
| 1 | 2 | 3 | 4 | 5 |
| 2 | 3 | 4 | 5 | 1 |
| 3 | 4 | 5 | 1 | 2 |
| 4 | 5 | 1 | 2 | 3 |
| 5 | 1 | 2 | 3 | 4 |

By interchanging the rows and columns we can get a number of Latin squares for 5 treatments.

Consider a Latin square of side r with treatments $T_1, \dots, T_r$. Let $y_{ij(k)}$ be the yield of the k-th treatment applied to the (i,j)-th plot. Assuming additivity of effect we may set up the model

$$y_{ijk} = g + \alpha_i + \beta_j + t_k + \varepsilon_{ij(k)}$$

where g is the effect common to all plots, $\alpha_i$ is the effect of the i-th row, $\beta_j$ is the effect of the j-th column, $t_k$ is the effect of the k-th treatment, and $\varepsilon_{ij(k)}$ is a random error assumed to be distributed normally with zero mean and variance $\sigma^2$. The total error S.S. is

$$L = \sum_{i,j,k} (y_{ijk} - g - \alpha_i - \beta_j - t_k)^2.$$

Then the least square normal equations are

$$G - r^2 g - r(\alpha_. + \beta_. - t_.) = 0, \tag{1}$$

$$R_i - rg - r\alpha_i - \beta_. - t_. = 0, \tag{2}$$

$$C_j - rg - \alpha_. - r\beta_j - t_. = 0, \tag{3}$$

$$T_k - rg - \alpha_. - \beta_. - rt_k = 0, \tag{4}$$

where G= grand total of all the observations, $R_i$ = $i$-th row total, $C_j$ = $j$-th column total, and $T_k$= sum of the observations corresponding to the $k$-th treatment.

There are $3r + 1$ equations to solve for the $3r + 1$ parameters. But

$$(1) = \sum_i (2) = \sum_j (3) = \sum_k (4).$$

Hence there are only $3r - 2$ independent equations. Introducing the subsidiary conditions $\alpha_. = \beta_. = t_. = 0$, the estimates will be

$$\hat{g} = \frac{G}{r^2}, \quad \hat{\alpha}_i = \frac{R_i}{r} - \frac{G}{r^2}, \quad \hat{\beta}_j = \frac{C_j}{r} - \frac{G}{r^2}, \quad \hat{t}_k = \frac{T_k}{r} - \frac{G}{r^2}.$$

Therefore the unconditional minimum is

$$S^2 = \sum_{i,j,k} (y_{ij(k)} - \hat{g} - \hat{\alpha}_i - \hat{\beta}_j - \hat{t}_k)^2$$

$$= \sum_{i,j,k} (y_{ij(k)} - \frac{R_i}{r} - \frac{C_j}{r} - \frac{T_k}{r} + \frac{2G}{r^2})^2$$

and it has got $r^2 - (3r - 2) = (r - 1)(r - 2)$ d.f..

The hypotheses we want to test are

(i) $H_0: t_1 = \cdots = t_r \ (= t, \text{say})$,
(ii) $H_1: \alpha_1 = \cdots = \alpha_r \ (= \alpha, \text{say})$,
(iii) $H_3: \beta_1 = \cdots = \beta_r \ (= \beta, \text{say})$,
each of rank $(r - 1)$.

Under $H_0$ the error S.S. is

$$L_0 = \sum_{i,j,k} (y_{ij(k)} - g - \alpha_i - \beta_j - t)^2.$$

The least square normal equations are

$$G - r^2 g - r(\alpha_. + \beta_.) - r^2 t = 0, \quad (5)$$

$$R_i - rg - r\alpha_i - \beta_. - rt = 0, \quad (6)$$

$$C_j - rg - \alpha_. - r\beta_j - rt = 0, \quad (7)$$

$$G - r^2 g - r(\alpha_. + \beta_.) - r^2 t = 0, \quad (8)$$

Consider the set of equations (5), (6) and (7). Out of the $(2r + 1)$ equations only $(2r - 1)$ are independent since $\sum_i(6) = \sum_j(7) = (5)$. Imposing the conditions $\alpha_. = 0, \beta_. = 0$, the estimate for $(g + t)$ is found to be

$$\widehat{(g + t)} = \frac{G}{r^2}.$$

From (6) and (7), the estimates of $\alpha_i$ and $\beta_j$ are

$$\hat{\alpha}_i = \frac{R_i}{r} - \frac{G}{r^2}, \quad \hat{\beta}_j = \frac{C_j}{r} - \frac{G}{r^2}.$$

Therefore the conditional minimum is

$$S_0^2 = \sum_{i,j} (y_{ij(k)} - \frac{R_i}{r} - \frac{C_j}{r} + \frac{G}{r^2})^2.$$

[Observe that in summing w.r.t. any one suffix, keeping fixed a second suffice, summation w.r.t. the third suffix also follows.]
Now,

$$S_0^2 = \sum_{i,j} (y_{ij(k)} - \frac{R_i}{r} - \frac{C_j}{r} - \frac{T_k}{r} + \frac{2G}{r^2} + \frac{T_k}{r} - \frac{G}{r^2})^2$$

$$= \sum_{i,j} (y_{ij(k)} - \frac{R_i}{r} - \frac{C_j}{r} - \frac{T_k}{r} + \frac{2G}{r^2})^2 + \sum_{i,j} (\frac{T_k}{r} - \frac{G}{r^2})^2,$$

since the product term vanishes. $S_0^2$ has $(r - 1)(r - 2) + (r - 1) = (r - 1)^2$ d.f.. Therefore S.S. due to treatments is

$$S_0^2 - S^2 = \sum_{i,j}\left(\frac{T_k}{r} - \frac{G}{r^2}\right)^2 = r\sum_k\left(\frac{T_k}{r} - \frac{G}{r^2}\right)^2.$$

Hence the test criterion for $H_0$ will be

$$F\big((r-1),(r-1)(r-2)\big) = \frac{(S_0^2 - S^2)/(r-1)}{S^2/(r-1)(r-2)}.$$

Similarly, the test criterion for $H_1$ and $H_2$ are

$$F\big((r-1),(r-1)(r-2)\big) = \frac{(S_1^2 - S^2)/(r-1)}{S^2/(r-1)(r-2)}$$

and

$$F\big((r-1),(r-1)(r-2)\big) = \frac{(S_2^2 - S^2)/(r-1)}{S^2/(r-1)(r-2)}$$

respectively, where

$$S_1^2 - S^2 = \sum_{i,j}\left(\frac{R_i}{r} - \frac{G}{r^2}\right)^2 = r\sum_i\left(\frac{R_i}{r} - \frac{G}{r^2}\right)^2$$

and

$$S_2^2 - S^2 = \sum_{i,j}\left(\frac{C_j}{r} - \frac{G}{r^2}\right)^2 = r\sum_j\left(\frac{C_j}{r} - \frac{G}{r^2}\right)^2.$$

We have

$$\text{Total S.S.} = \sum_{i,j}(y_{ij(k)} - \frac{G}{r^2})^2 = \sum_{i,j} y_{ij(k)}^2 - \frac{G^2}{r^2},$$

$$\text{Row S.S.} = r\sum_i\left(\frac{R_i}{r} - \frac{G}{r^2}\right)^2 = \sum_i \frac{R_i^2}{r} - \frac{G^2}{r^2},$$

$$\text{Column S.S.} = r\sum_j\left(\frac{C_j}{r} - \frac{G}{r^2}\right)^2 = \sum_j \frac{C_j^2}{r} - \frac{G^2}{r^2},$$

$$\text{Treatment S.S.} = r\sum_k\left(\frac{T_k}{r} - \frac{G}{r^2}\right)^2 = \sum_k \frac{T_k^2}{r} - \frac{G^2}{r^2}.$$

**Analysis of Variance Table**

| Source | S.S. | d.f. | M.S.S. | F |
|---|---|---|---|---|
| Rows | $\sum_i \frac{R_i^2}{r} - \frac{G^2}{r^2}$ | $r-1$ | $V_R$ | $F_1 = V_R/V_E$ |
| Columns | $\sum_j \frac{C_j^2}{r} - \frac{G^2}{r^2}$ | $r-1$ | $V_C$ | $F_2 = V_C/V_E$ |
| Treatments | $\sum_k \frac{T_k^2}{r} - \frac{G^2}{r^2}$ | $r-1$ | $V_T$ | $F_3 = V_T/V_E$ |
| Error | (Sub) | (Sub) | $V_E$ | |
| Total | $\sum_{i,j} y_{ij(k)}^2 - \frac{G^2}{r^2}$ | $r^2-1$ | | |

## 3.20 Judging the Relative Merits of Treatments

If the test above shows a significant deviation in treatment effects, we test hypotheses of the form $t_p = t_q$ to judge the relative merits of treatments.

$$\widehat{(t_p - t_q)} = \frac{T_p - T_q}{r} \quad \text{with} \quad V(\hat{t}_p - \hat{t}_q) = \frac{2}{r}\sigma^2.$$

Best unbiased estimate of $\sigma^2 = V_E$. Hence

$$\frac{\frac{T_p - T_q}{r}}{\sqrt{\frac{2}{r}V_E}}$$

will be a Students' t with $(r-1)(r-2)$ d.f. and can be used to test the difference $t_p - t_q$.

Let $t_0$ be the tabled value of t at 5% level. Then the treatments $t_p$ and $t_q$ are significantly different if

$$\left|\frac{T_p - T_q}{r}\right| \geq t_0 \sqrt{\frac{2}{r}V_E}.$$

Similarly, $\frac{T_p - T_q}{r} \pm t_0 \sqrt{\frac{2}{r} V_E}$ will be 95% confidence limits for the difference $t_p - t_q$.

### 3.21 Graeco Latin Square

A mutually orthogonal Latin square of two sets of treatments is called a Graeco Latin square. In an ordinary Latin square all the rows as well as all the columns should contain all the treatments. In the Graeco Latin square, in every row as well as in every column we will be taking all the treatments in the two sets and every member of one set will occur once and only once with every member of the other set. This design also achieves a two-way elimination of heterogeneity and it is mainly used when we want to compare two sets of treatments.

Consider a Graeco Latin square of side r. Let $T_1, \dots, T_r$ and $\Gamma_1, \dots, \Gamma_r$ be the two sets of treatments. Let $y_{ij(kl)}$ be the observation corresponding to the treatments $T_k$ and $\Gamma_l$ when applied in the $j$-th plot of the $i$-th block.

Let $\alpha_i$ and $\beta_j$ be the row and column effects. Let $t_k$ denote the effect due to treatment $T_k$ and $\gamma_l$ denote the effect due to treatment $\Gamma_l$. The general effect may be combined with the row and column effects so that assuming additivity of effects the model may be written

$$y_{ij(kl)} = \alpha_i + \beta_j + t_k + \gamma_l + \varepsilon_{ij}$$

where $\varepsilon_{ij}$ is random error assumed distributed normally around zero with common variance $\sigma^2$. The total error S.S. in

$$L = \sum_{i,j} (y_{ij(kl)} - \alpha_i - \beta_j - t_k - \gamma_l)^2.$$

The least square normal equations are

$$R_i - r\alpha_i - \beta_. - t_. - \gamma_. = 0, \quad (1)$$

$$C_j - \alpha_. - r\beta_j - t_. - \gamma_. = 0, \quad (2)$$

$$T_k - \alpha_. - \beta_. - rt_k - \gamma_. = 0, \quad (3)$$

$$\Gamma_l - \alpha_. - \beta_. - t_. - r\gamma_l = 0, \quad (4)$$

where $T_k$ ($\Gamma_l$) is the sum of all the yields corresponding to treatment $T_k$ ($\Gamma_l$).

There are 4r equations. But since $\sum_i(1) = \sum_j(2) = \sum_k(3) = \sum_l(4)$, there are only $4r - 3$ independent equations. Introducing the subsidiary condition $\alpha_. = \beta_. = t_. = 0$, the solutions are found to be

$$\hat{\gamma}_l = \frac{\Gamma_l}{r}, \ \hat{t}_k = \frac{T_k}{r} - \frac{G}{r^2}, \ \hat{\beta}_j = \frac{C_j}{r} - \frac{G}{r^2}, \ \hat{\alpha}_i = \frac{R_i}{r} - \frac{G}{r^2}.$$

Therefore the unconditional minimum S.S. is

$$S^2 = \sum_{i,j} (y_{ij(kl)} - \frac{R_i}{r} - \frac{C_j}{r} - \frac{T_k}{r} - \frac{\Gamma_l}{r} + \frac{3G}{r^2})^2,$$

and this has $r^2 - (4r - 3) = (r - 1)(r - 3)$ d.f..

The hypotheses we wish to test are

$$H_0: \alpha_1 = \cdots = \alpha_r \ (= \alpha, \text{say}),$$

$$H_1: \beta_1 = \cdots = \beta_r \ (= \beta, \text{say}),$$

$$H_2: t_1 = \cdots = t_r \ (= t, \text{say}),$$

$$H_3: \gamma_1 = \cdots = \gamma_r \ (= \gamma, \text{say}),$$

each of rank $(r - 1)$.

Under $H_0$, the total error S.S. is

$$L_0 = \sum_{i,j} (y_{ij(k)} - \alpha - \beta_j - t_k - \gamma_l)^2.$$

The least square normal equations are

$$G - r^2\alpha - r\beta_. - rt_. - r\gamma_. = 0 \qquad (5)$$

$$C_j - r\alpha - r\beta_j - t_. - \gamma_. = 0, \qquad (6)$$

$$T_k - r\alpha - \beta_. - rt_k - \gamma_. = 0, \qquad (7)$$

$$\Gamma_l - r\alpha - \beta_. - t_. - r\gamma_l = 0, \qquad (8)$$

of which $(3r - 2)$ are independent. Imposing the conditions $\beta_. = \gamma_. = t_. = 0$, the solutions are

$$\hat{\alpha} = \frac{G}{r^2}, \ \hat{\beta}_j = \frac{C_j}{r} - \frac{G}{r^2}, \ \hat{t}_k = \frac{T_k}{r} - \frac{G}{r^2}, \ \hat{\gamma}_l = \frac{\Gamma_l}{r} - \frac{G}{r^2},$$

and the conditional minimum S.S. is

$$S_0^2 = \sum_{i,j}(y_{ij(kl)} - \frac{R_i}{r} - \frac{C_j}{r} - \frac{T_k}{r} - \frac{\Gamma_l}{r} + \frac{3G}{r^2})^2 + \sum_{i,j}(\frac{R_i}{r} - \frac{G}{r^2})^2,$$

with $(r-1)(r-3) + (r-1) = (r-1)(r-2)$ d.f..

Therefore the S.S. due to row effects is

$$S_0^2 - S^2 = \sum_{i,j}\left(\frac{R_i}{r} - \frac{G}{r^2}\right)^2 = r\sum_{i}(\frac{R_i}{r} - \frac{G}{r^2})^2$$

with $(r-1)$ d.f. and hence the test criterion to test $H_0$ is

$$F\big((r-1),(r-1)(r-3)\big) = \frac{(S_0^2 - S^2)/(r-1)}{S^2/(r-1)(r-3)}.$$

By an exactly similar procedure we can get test criterion to test $H_1$, $H_2$ and $H_3$. The analysis of variance is carried out as in the table below.

| Source | S.S. | d.f. | M.S.S. | F |
|---|---|---|---|---|
| Rows | $\sum_i \frac{R_i^2}{r} - \frac{G^2}{r^2}$ | $r-1$ | $V_R$ | $F_0 = V_R/V_E$ |
| Columns | $\sum_j \frac{C_j^2}{r} - \frac{G^2}{r^2}$ | $r-1$ | $V_C$ | $F_1 = V_C/V_E$ |
| Treatments (First set) | $\sum_k \frac{T_k^2}{r} - \frac{G^2}{r^2}$ | $r-1$ | $V_T$ | $F_2 = V_T/V_E$ |
| Treatments (Second set) | $\sum_l \frac{\Gamma_l^2}{r} - \frac{G^2}{r^2}$ | $r-1$ | $V_\Gamma$ | $F_3 = V_\Gamma/V_E$ |
| Error | (Sub) | (Sub) | $V_E$ | |
| Total | $\sum_{i,j} y_{ij(kl)}^2 - \frac{G^2}{r^2}$ | $r^2-1$ | | |

**Judging relative merits of treatments**

If $F_2$ is significant we can test individual hypotheses such as $t_i = t_j$. The criterion will be

$$t = \frac{\frac{T_i - T_j}{r}}{\sqrt{\frac{2}{r} V_E}}$$

which is a Students' t with $(r-1)(r-3)$ d.f..

Similarly, if $F_3$ is significant, the criterion to test the hypothesis $\gamma_p = \gamma_q$ is

$$t = \frac{\frac{\Gamma_p - \Gamma_q}{r}}{\sqrt{\frac{2}{r} V_E}}$$

which is also a Students' t with $(r-1)(r-3)$ d.f. .

## 3.22 Hyper – Graeco Latin Square

Suppose we want to compose p sets of r treatments each $(p < r)$. There exists a number of Latin squares of side r. Consider an orthogonal Latin square of sider r such that in every row and every column we will get all the treatments and every treatment of one set occurs with every treatment of another set once and only once. [The maximum number of orthogonal Latin square of side r is $(r-1)$. Since $p < r$, this Latin square exists.] Such a Latin square is known as a hyper – Graeco Latin square. The analysis is just an extension of the analysis of a Graeco Latin square design.

# Chapter 4

# Analysis of Covariance (ANCOVA)

Throughout the previous chapter it was assumed that the mean value of the random variable in question consists of the sum of a general effect (which is the same for all random variables) and effects due to rows, columns, treatments, interaction, etc. It frequently happens that there are practical situations which suggest that the mean value of the random variable should include linear functions of one or more fixed variates in addition to the sum of effects of the type mentioned above. For instance, if $y_{ij}$ refers to yield of wheat in the $(i,j)$-th plot of a two-way layout, not only should the mean value of $y_{ij}$ include a general effect and row and column effects, but also linear effect of the number of plants in this plot, say $x_{ij}$. The mean value would then be given by

$$E(y_{ij}) = \alpha_i + \beta_j + a + bx_{ij}\,, \quad j = 1, \dots, n_i; \quad i = 1, \dots, k$$

where $\alpha_i$ is the i-th row effect, $\beta_j$ is the j-th column effect, a is the general effect, and b is the coefficient of regression.

The object of the analysis of covariance is to examine the modifications that should be made in the ordinary analysis of variance in order to take one or more fixed variates into account in the mean value of the random variable in question. The additional variables taken into consideration are called concomitant variables.

## 4.1 One-Way Classification with a Single Concomitant Variable

Let there be k samples of sizes $n_1, , \dots, n_k$ from k populations with group effects $\alpha_1, \dots, \alpha_k$. Let $y_{ij}$ be the j-th observation in the i-th sample. Let $x_{ij}$ be the corresponding observation on the concomitant variable. Let us assume that there is a linear regression of the form $a + bx_{ij}$ of $y_{ij}$ on $x_{ij}$. Then the model will be

$$y_{ij} = \alpha_i + a + bx_{ij} + \varepsilon_{ij}, \quad j = 1, \ldots, n_i; \quad i = 1, \ldots, k$$

where $\varepsilon_{ij}$, as usual, is a random error assumed to be distributed normally around zero with a common variance $\sigma^2$. The total error S. S. is

$$L = \sum_{i,j} (y_{ij} - \alpha_i - a - bx_{ij})^2.$$

The least square normal equations are

$$y_{i.} - n_i\alpha_i - n_i a - bx_{i.} = 0 \tag{1}$$

$$y_{..} - \sum n_i\alpha_i - n_. a - bx_{..} = 0 \tag{2}$$

$$\sum_{i,j} x_{ij}\, y_{ij} - \sum_i \alpha_i x_{i.} - ax_. - b\sum_{i,j} x_{ij}^2 = 0 \tag{3}$$

Altogether there are $(k+2)$ equations. But since $\sum_i(1) = (2)$, only $(k+1)$ equations are independent. Imposing the condition $a = 0$, the estimates are

$$\hat{\alpha}_i = \frac{y_{i.}}{n_i} - \hat{b}\frac{x_{i.}}{n_i}, \qquad \hat{b} = \frac{\sum_{i,j} x_{ij}y_{ij} - \sum_i \frac{x_{i.}y_{i.}}{n_i}}{\sum_{i,j} x_{ij}^2 - \sum_i \frac{x_{i.}^2}{n_i}}.$$

Letting

$$E_{xy} = \sum_{i,j} x_{ij}y_{ij} - \sum_i \frac{x_{i.}y_{i.}}{n_i}, \qquad E_x = \sum_{i,j} x_{ij}^2 - \sum_i \frac{x_{i.}^2}{n_i},$$

$$E_y = \sum_{i,j} y_{ij}^2 - \sum_i \frac{y_{i.}^2}{n_i}.$$

We have $\hat{b} = \frac{E_{xy}}{E_x}$.

The unconditional minimum S.S. is

$$S^2 = \sum_{i,j} (y_{ij} - \hat{\alpha}_i - \hat{b}x_{ij})^2$$
$$= \sum_{i,j} (y_{ij} - \hat{\alpha}_i - \hat{b}x_{ij})y_{ij},$$

by the normal equations

$$= \left(\sum_{i,j} y_{ij}^2 - \sum_i \frac{y_{i.}^2}{n_i}\right)$$
$$- \hat{b}\left(\sum_{i,j} x_{ij}y_{ij} - \sum_i \frac{x_{i.}y_{i.}}{n_i}\right)$$
$$= E_y - \hat{b}E_{xy}$$
$$= E_y - \frac{(E_{xy})^2}{E_x}.$$

The d.f. of $S^2$ are $n_. - k - 1$.

**(i) Consider $H_0$:** $b = 0$

Then

$$L_0 = \sum_{i,j} (y_{ij} - \alpha_i - a)^2.$$

The least square estimate of $\alpha_i + a$ is

$$(\widehat{\alpha_i + a}) = \frac{y_{i.}}{n_i}.$$

Therefore the conditional minimum is

$$S_0^2 = \sum_{i,j} \left(y_{ij} - \frac{y_{i.}}{n_i}\right)^2$$
$$= E_y, \text{ with } (n_. - k) \text{ d.f.}$$

So the test criterion for regression is

$$F(1, n_{.} - k - 1) = \frac{(S_0^2 - S^2)/1}{S^2/(n_{.} - k - 1)}.$$

**(ii) Consider again $H_1$: $\alpha_1 = \alpha_2 = \cdots = \alpha_k$**

Under the assumption that there is regression, the observational equations will be

$$E(y_{ij}) = \alpha + a + bx_{ij}$$

where $\alpha$ is the common group effect. So the total error S.S. is

$$L_1 = \sum_{i,j} (y_{ij} - \alpha - a - bx_{ij})^2.$$

The usual least square procedure gives

$$\widehat{(\alpha + a)} = \frac{y_{..}}{n_{.}} - \hat{b}\frac{x_{..}}{n_{.}}, \qquad \hat{b} = \frac{\sum_{i,j} x_{ij}y_{ij} - \sum_i \frac{x_{..}y_{..}}{n_{.}}}{\sum_{i,j} x_{ij}^2 - \frac{x_{..}^2}{n_{.}}} = \frac{T_{xy}}{T_x}$$

where $T_{xy} = \sum_{i,j} x_{ij}y_{ij} - \sum_i \frac{x_{..}y_{..}}{n_{.}}$, $T_x = \sum_{i,j} x_{ij}^2 - \frac{x_{..}^2}{n_{.}}$. Therefore the minimum S.S. under $H_1$ is $S_1^2 = T_y - \frac{T_{xy}}{T_x}$, on simplification, with $(n_{.} - 1)$ d.f.. The S.S. due to group effect is

$$S_1^2 - S^2 = (T_y - E_y) - \left(\frac{T_{xy}^2}{T_x} - \frac{E_{xy}^2}{E_x}\right) \text{ with } (k-1) \text{ d.f..}$$

So the test criterion for $H_1$ is

$$F(k-1, n_{.} - k - 1) = \frac{(S_1^2 - S^2)/(k-1)}{S^2/(n. - k - 1)}.$$

Letting

$$B_{xy} = \sum_i \frac{x_{i.}y_{i.}}{n_i} - \frac{x_{..}y_{..}}{n_{.}}, \qquad B_x = \sum_i \frac{x_{i.}^2}{n_i} - \frac{x_{..}^2}{n_{.}},$$

we have $T_x = E_x + B_x$, $T_y = E_y + B_y$ and $T_{xy} = E_{xy} + B_{xy}$.

**Analysis of Covariance Table**

| Source | S.S. and Products | d.f. | Ressidual | d.f. | M.S.S. | F |
|---|---|---|---|---|---|---|
| | $y$ $x$ $xy$ | | | | | |
| Between class | $B_y$ $B_x$ $B_{xy}$ | $k-1$ | (Sub) | $k-1$ | $V_A$ | $\frac{V_A}{V_E}$ |
| Within class | $E_y$ $E_x$ $E_{xy}$ | $n_{.}-k$ | $E_y-\frac{(E_{xy})^2}{E_x}$ | $n_{.}-k-1$ | $V_E$ | |
| Total | $T_y$ $T_x$ $T_{xy}$ | $n_{.}-1$ | $T_y-\frac{(T_{xy})^2}{T_x}$ | $n_{.}-2$ | | |

The F-ratio $\frac{V_A}{V_E}$ will test the significance of the deviation in α-effects. If this F is significant, then we cannot assume the equality of the α-effects and we proceed to test individual hypotheses such as $\alpha_i = \alpha_j$ using the t-test, even though the test has so many limitations.

$$\hat{\alpha}_i - \hat{\alpha}_j = (\bar{y}_{i.} - \bar{y}_{j.}) - \frac{E_{xy}}{E_x}(\bar{x}_{i.} - \bar{x}_{j.})$$

$$V(\hat{\alpha}_i - \hat{\alpha}_j) = \left[\left(\frac{1}{n_i} - \frac{1}{n_j}\right) + \frac{(\bar{x}_{i.} - \bar{x}_{j.})^2}{E_x}\right]\sigma^2,$$

and $\sigma^2$ is best estimated by $V_E$ which has $(n_{.} - k - 1)$ d.f.. So the Student's t will be

$$t = \frac{(\bar{y}_{i.} - \bar{y}_{j.}) - \frac{E_{xy}}{E_x}(\bar{x}_{i.} - \bar{x}_{j.})}{\sqrt{\left[\left(\frac{1}{n_i} - \frac{1}{n_j}\right) + \frac{(\bar{x}_{i.} - \bar{x}_{j.})^2}{E_x}\right]V_E}} \quad \text{with } (n_{.} - k - 1) \text{ d.f.}\ .$$

## 4.2 Two-Way Classification with One Observation per Cell and a Single Concomitant Variable

Assuming additivity of effects and linearity of regression, the model will be

$$y_{ij} = \alpha_i + \beta_j + a + bx_{ij} + \varepsilon_{ij}\,, \quad i = 1, \ldots, k; \quad j = 1, \ldots, q$$

with usual assumptions about $\varepsilon_{ij}$. Absorbing a in $\alpha_i$ or $\beta_j$, the observational equations will be

$$E(y_{ij}) = \alpha_i + \beta_j + bx_{ij}\,.$$

(We assume there is no correlation between the concomitant variable and the block or column effects.) The least square normal equations are

$$y_{i.} - q\alpha_i - \beta_. - bx_{i.} = 0\,, \tag{4}$$

$$y_{.j} - \alpha_. - p\beta_j - bx_{.j} = 0, \tag{5}$$

$$\sum_{i,j} x_{ij}y_{ij} - \sum_i \alpha_i x_{i.} - \sum_j \beta_j x_{.j} - b\sum_{i,j} x_{ij}^2 = 0. \tag{6}$$

There are $(p+q+1)$ equations to estimate the $(p+q+1)$ parameters. But, since $\Sigma_i(4) = \Sigma_j(5)$, we may impose one condition. Let it be $\beta_. = 0$. Then

$$\hat{\alpha}_i = \frac{y_{i.}}{q} - \hat{b}\frac{x_{i.}}{q}, \qquad \hat{\beta}_j = \left(\frac{y_{.j}}{q} - \frac{y_{..}}{pq}\right) - \hat{b}\left(\frac{x_{.j}}{p} - \frac{x_{..}}{pq}\right).$$

Substituting these in (6), and simplifying, we get

$$\hat{b} = \frac{E_{xy}}{E_x}$$

where

$$E_{xy} = \sum_{i,j} x_{ij}y_{ij} - \sum_i \frac{x_{i.}y_{i.}}{q} - \sum_j \frac{x_{.j}y_{.j}}{p} + \frac{x_{..}y_{..}}{pq},$$

$$E_x = \sum_{i,j} x_{ij}^2 - \sum_i \frac{x_{i.}^2}{q} - \sum_j \frac{x_{.j}^2}{p} + \frac{x_{..}^2}{pq}.$$

Therefore the unconditional minimum is

$$S^2 = E_y - \hat{b}E_{xy}\text{, (on necessary subistition)}$$
$$= E_y - \frac{(E_{xy})^2}{E_x} \quad \text{(and simplification)}$$

with $(pq - p - q)$ d.f..

**(i) Consider the hypothesis $H_0$:** $b = 0$

As usual, setting up the normal equations and putting $\beta_. = 0$, we get

$$\hat{\alpha}_i = \frac{y_{i.}}{q}, \qquad \hat{\beta}_j = \frac{y_{.j}}{q} - \frac{y_{..}}{pq}.$$

So the conditional minimum is $S_0^2 = E_y$, (on simplification,) $S_0^2$ has $pq - p - q + 1 = (p-1)(q-1)$ d.f. . Therefore S.S. due to regression is $S_0^2 - S^2 = \frac{(E_{xy})^2}{E_x}$ with one d. f. and the F-ratio is

$$F(1, pq - p - q) = \frac{S_0^2 - S^2}{S^2/(pq - p - q)}.$$

**(ii) Consider the hypothesis $H_1$:** $\alpha_1 = \alpha_2 = \cdots = \alpha_p (= \alpha$, **say)**

The least square normal equations under $H_1$ are

$$y_{..} - pq\alpha - p\beta_. - bx_{..} = 0 \tag{7}$$
$$y_{.j} - p(\alpha + \beta_j) - bx_{.j} = 0 \tag{8}$$
$$\sum_{i,j} x_{ij}y_{ij} - \alpha x_{..} - \sum_j \beta_j x_{.j} - b\sum_{i,j} x_{ij}^2 = 0 \tag{9}$$

Imposing the condition $\beta_. = 0$, (7) gives

$$\hat{\alpha} = \frac{y_{..}}{pq} - \hat{b}\frac{x_{..}}{pq},$$

and (8) then gives

$$\hat{\beta}_j = \left(\frac{y_{.j}}{q} - \frac{y_{..}}{pq}\right) - \hat{b}\left(\frac{x_{.j}}{p} - \frac{x_{..}}{pq}\right).$$

Substituting in (9) and solving, we get

$$\hat{b} = \frac{E_{xy} + B_{xy}}{E_x + B_x},$$

where

$$B_{xy} = \sum_i \frac{x_{i.}y_{i.}}{q} - \frac{x_{..}y_{..}}{pq}, \qquad B_x = \sum_i \frac{x_{i.}^2}{q} - \frac{x_{..}^2}{pq}.$$

Therefore the conditional minimum S.S. is

$$S_1^2 = (E_y + B_y) - \frac{(E_{xy} + B_{xy})^2}{E_x + B_x},$$

and the F-ratio to test block effects is

$$F(p-1, pq-p-q) = \frac{(S_0^2 - S^2)/(p-1)}{S^2/(pq-p-q)}.$$

**(iii) Consider the hypothesis $H_2$: $\beta_1 = \beta_2 = \cdots = \beta_q = \beta$**

Similar to (ii) above, the F-ratio to test for treatment effects is

$$F(q-1, pq-p-q) = \frac{(S_2^2 - S^2)/(q-1)}{S^2/(pq-p-q)}$$

where

$$S_2^2 = (E_y + T_y) - \frac{(E_{xy} + T_{xy})^2}{E_x + T_x}$$

where

$$T_{xy} = \sum_j \frac{x_{.j}y_{.j}}{p} - \frac{x_{..}y_{..}}{pq}, \qquad T_x = \sum_j \frac{x_{.j}^2}{p} - \frac{x_{..}^2}{pq}.$$

**Analysis of covariance table** (broken into two tables)

| Source | SS y | and x | Product xy | d.f. | Residual |
|---|---|---|---|---|---|
| Blocks<br>Treatments<br>Error | $B_y$<br>$T_y$<br>$E_y$ | $B_x$<br>$T_x$<br>$E_x$ | $B_{xy}$<br>$T_{xy}$<br>$E_{xy}$ | $p-1$<br>$q-1$<br>$(p-1)(q-1)$ | $a = A - C$<br>$b = B - C$<br>$E_y - \frac{(E_{xy})^2}{E_x} = C$ |
| Total | $\sum_{i.j} y_{ij}^2 - \frac{y_{..}^2}{pq}$ | $\sum_{i.j} x_{ij}^2 - \frac{x_{..}^2}{pq}$ | $\sum_{i.j} x_{ij}y_{ij} - \frac{x_{..}y_{..}}{pq}$ | $pq-1$ | |
| Blocks+Error | $B_y + E_y$ | $B_x + E_x + E_{xy}$ | $B_{xy}$ | $q(p-1)$ | $A = (B_y + E_y) - \frac{(B_{xy} + E_{xy})^2}{B_x + E_x}$ |
| Treatment +Error | $T_y + E_y$ | $T_x + E_x + E_{xy}$ | $T_{xy}$ | $p(q-1)$ | $B = (T_y + E_y) - \frac{(T_{xy} + E_{xy})^2}{T_x + E_x}$ |

| Source | d.f. | M.S.S | F |
|---|---|---|---|
| Blocks<br>Treatments<br>Error | $p-1 = a'$<br>$q-1 = b'$<br>$(p-1)(q-1) - 1 = c'$ | $V_B = \frac{a}{a'}$<br>$V_T = \frac{b}{b'}$<br>$V_E = \frac{c}{c'}$ | $F_1 = \frac{V_B}{V_E}$<br>$F_2 = \frac{V_T}{V_E}$ |
| Total | | | |
| Blocks+Error | $q(p-1)$ | | |
| Treatment +Error | $p(q-1) - 1$ | | |

**(iv) If F in (ii) and F in (iii) are significant, we proceed to test individual hypotheses such as** $\alpha_i = \alpha_j$ **or** $\beta_l = \beta_k$

We have

$$\hat{\beta}_l - \hat{\beta}_k = (\bar{y}_{.l} - \bar{y}_{.k}) - \frac{E_{xy}}{E_x}(\bar{x}_{.l} - \bar{x}_{.k})$$

and

$$V(\hat{\beta}_l - \hat{\beta}_k) = \left[\frac{2}{p} + \frac{(\bar{x}_{.l} - \bar{x}_{.k})^2}{E_x}\right]\sigma^2$$

where $\sigma^2$ is best estimated by $V_E$ with $(pq - p - q)$ d.f.. So the Student's t for testing $\beta_l = \beta_k$ is

$$t = \frac{(\bar{y}_{.l} - \bar{y}_{.k}) - \frac{E_{xy}}{E_x}(\bar{x}_{.l} - \bar{x}_{.k})}{\sqrt{\left[\frac{2}{p} + \frac{(\bar{x}_{.l} - \bar{x}_{.k})^2}{E_x}\right]V_E}}.$$

with $(pq - p - q)$ d.f..

The technique can be generalized to the case when there are k concomitant variables.

## 4.3 Some Examples of ANCOVA

### (i) Randomised Blocks

The following table gives the plan and yield in ounces from a Randomised Block experiment with six varieties of paddy seeds tried out in five blocks. Figures in brackets indicate the variety number

| | | | | | | |
|---|---|---|---|---|---|---|
| Block I | (3)<br>43 | (1)<br>14 | (4)<br>22 | (6)<br>19 | (5)<br>19 | (2)<br>16 |
| Block II | (2)<br>31 | (6)<br>19 | (3)<br>40 | (1)<br>17 | (4)<br>15 | (5)<br>13 |
| Block III | (5)<br>19 | (2)<br>20 | (6)<br>13 | (4)<br>20 | (1)<br>20 | (3)<br>23 |
| Block IV | (1)<br>15 | (4)<br>17 | (2)<br>26 | (3)<br>29 | (5)<br>25 | (6)<br>14 |
| Block V | (6)<br>19 | (3)<br>30 | (5)<br>15 | (2)<br>19 | (4)<br>15 | (1)<br>21 |

For convenience of analysis we rearrange the table as follows:

| Treatments / Blocks | (1) | (2) | (3) | (4) | (5) | (6) | Block Totals $y_{i.}$ |
|---|---|---|---|---|---|---|---|
| I | 14 | 16 | 43 | 22 | 19 | 19 | 133 |
| I | 17 | 31 | 40 | 15 | 13 | 19 | 135 |
| III | 20 | 20 | 23 | 20 | 19 | 13 | 115 |
| IV | 15 | 26 | 29 | 17 | 25 | 14 | 126 |
| V | 21 | 19 | 30 | 15 | 15 | 19 | 119 |
| Variety Total $y_{.j}$ | 87 | 112 | 165 | 89 | 91 | 84 | 628 |

After necessary computations, the analysis of variance table is set up as follows:

**Analysis of Variance Table**

| Source | S.S. | d. f. | M.S.S. | F | Tabled Value | Inference |
|---|---|---|---|---|---|---|
| Block | 49.87 | 4 | 12.47 | 0.44 | 2.87 | Not significant |
| Variety | 973.07 | 5 | 194.61 | 6.89 | 2.71 | Highly significant |
| Error | 566.93 | 20 | 28.35 | | | |
| Total | 1589.87 | 29 | | | | |

Mean of varieties in descending order of preference is given below

| Variety | 3 | 2 | 5 | 4 | 1 | 6 |
|---|---|---|---|---|---|---|
| Average yield | 33.0 | 22.4 | 18.2 | 17.8 | 17.4 | 16.3 |

**(ii) Latin Square**

The following table gives the plan and yield in pounds for a Latin square design conducted with five varieties of Tapioca. The figures in brackets give the variety number.

| Rows \ Columns | 1 | 2 | 3 | 4 | 5 | Row Total $R_i$ |
|---|---|---|---|---|---|---|
| 1 | (1) 47 | (5) 48 | (4) 37 | (3) 44 | (2) 48 | 224 |
| 2 | (4) 54 | (2) 58 | (1) 37 | (5) 51 | (3) 45 | 245 |
| 3 | (2) 42 | (1) 18 | (3) 27 | (4) 37 | (5) 47 | 171 |
| 4 | (3) 22 | (4) 26 | (5) 37 | (2) 39 | (1) 29 | 153 |
| 5 | (5) 32 | (3) 23 | (2) 22 | (1) 23 | (4) 23 | 123 |
| Column Total $C_j$ | 197 | 173 | 150 | 204 | 192 | 916 |

The treatment totals are given below:

| Treatment No. | 1 | 2 | 3 | 4 | 5 |
|---|---|---|---|---|---|
| Treatment Total $T_k$ | 154 | 209 | 161 | 177 | 215 |

**Analysis of Variance Table**

| Source | S.S. | d.f. | M.S.S. | F | Tabled Value | Inference |
|---|---|---|---|---|---|---|
| Rows | 2033.76 | 4 | 508.44 | 32.36 | 3.26 | Significant |
| Columns | 381.36 | 4 | 95.34 | 6.36 | 3.26 | Significant |
| Treatments | 612.16 | 4 | 153.04 | 9.74 | 3.26 | Significant |
| Error | 188.48 | 12 | 15.71 | | | |
| Total | 3215.76 | 24 | | | | |

Average yield of treatments in descending order is given below.

| Treatment | 5 | 2 | 4 | 3 | 1 |
|---|---|---|---|---|---|
| Average Yield | 43.0 | 41.8 | 35.4 | 32.2 | 30.8 |

The critical difference is $t_0 \times \sqrt{\frac{2}{r}V_E} = 2.179 \times \sqrt{6.284} = 5.46$. This shows that the variety numbers 5 and 2 are superior to the other three even though there is nothing to choose between them.

# Chapter 5

# Missing and Mixed Plots

## 5.1 Two-Way Classification with One Cell Missing

Let there be p blocks and q treatments. Let the observation in the (i,j)-th cell be missing. Let x denote the missing observation. Performing the ordinary analysis of variance, we get the following table:

| Variation due to | S.S. |
|---|---|
| Blocks | $\frac{y_{1.}^2}{q}+\cdots+\frac{(y_{i.}'+x)^2}{q}+\cdots+\frac{y_{p.}^2}{q}-\frac{(y_{..}'+x)^2}{pq}$ |
| Treatments | $\frac{y_{.1}^2}{p}+\cdots+\frac{(y_{.j}'+x)^2}{p}+\cdots+\frac{y_{.p}^2}{p}-\frac{(y_{..}'+x)^2}{pq}$ |
| Error | (By subtraction) |
| Total | $y_{11}^2+\cdots+x^2+\cdots+y_{pq}^2-\frac{(y_{..}'+x)^2}{pq}$ |

In this table $y_{i.}'$ and $y_{.j}'$ denote the sum of the existing observations in the i-th row and j-th column respectively and $y_{..}'$ denotes the grand total of all the existing observations. The error S.S. is

$$E = x^2 - \frac{(y_{i.}'+x)^2}{q} - \frac{(y_{.j}'+x)^2}{p} + \frac{(y_{..}'+x)^2}{pq}$$
$$+ \text{ terms independent of } x.$$

We estimate x by minimizing E w.r.t. x. Then the estimating equation is

$$x - \frac{y_{i.}'+x}{q} - \frac{y_{.j}'+x}{p} + \frac{y_{..}'+x}{pq} = 0$$

giving

$$x = \frac{py'_{i.} + qy'_{.j} - y'_{..}}{(p-1)(q-1)}.$$

The same procedure applies in the case of a randomized block also.

Let the block and treatment effects be $\alpha_1, \ldots, \alpha_p$ and $\beta_1, \ldots, \beta_q$. Suppose we want to test the contrast $\alpha_i - \alpha_l$ when the (i,j)-th observation is missing. Then

$$\begin{aligned}\hat{\alpha}_i - \hat{\alpha}_l &= \frac{y'_{i.} + x}{q} - \frac{y_{l.}}{q} \\ &= \frac{y'_{i.}}{q} + \frac{py'_{i.} + qy'_{.j} - y'_{..}}{q(p-1)(q-1)} - \frac{y_{l.}}{q}.\end{aligned}$$

$$\begin{aligned}V(\hat{\alpha}_i - \hat{\alpha}_l) &= \left[\frac{1}{q} + \frac{p}{q(p-1)(q-1)} + \frac{1}{q}\right]\sigma^2 \\ &= \left[\frac{2}{q} + \frac{p}{q(p-1)(q-1)}\right]\sigma^2.\end{aligned}$$

When there is no missing value the variance is $\frac{2}{q}\sigma^2$.

A similar procedure gives

$$V(\hat{\beta}_j - \hat{\beta}_l) = \left[\frac{2}{p} + \frac{q}{p(p-1)(q-1)}\right]\sigma^2.$$

The estimate of x can be inserted in the original table and the analysis of the augmented table can be done as usual.

**Analysis-Modification**

An approximate test of significance of the null hypothesis that the treatments have no differential effects may be obtained by analyzing the augmented table in the usual way, with the modification that the d. f. for the error S.S. is reduced by the member of missing plots (in this case 1). In this case the analysis of variance table is called the Completed Table.

**Completed Table**

| Variation due to | S.S. | d.f. | M.S.S. | F |
|---|---|---|---|---|
| Blocks | $\sum_i \frac{y_{i.}^2}{q} - \frac{y_{..}^2}{pq}$ | p-1 | $V_B$ | $F_1$ |
| Treatments | $\sum_i \frac{y_{.j}^2}{q} - \frac{y_{..}^2}{pq}$ | q-1 | $V_T$ | $F_2$ |
| Error | (Sub) | (sub) | $V_E$ | |
| Total | $\sum_{i,j} y_{ij}^2 - \frac{y_{..}^2}{pq}$ | pq-2 | | |

If $F_2$ is not significant we assume the hypothesis $\beta_1 = \beta_2 = \cdots = \beta_q$ to be true.

This test can be shown to be biased in that the expectation of the treatment mean square is greater than the expectation of the error mean square under the null hypothesis. If the approximate test of significance indicates that there are no significant treatment differences, there is no need to perform the accurate test of significance.

Therefore when $F_2$ is significant the accurate test of significance is made by the analysis of variance table below.

**Analysis of variance table for exact test of significance**

| Variation due to | S. S. | d. f. |
|---|---|---|
| Blocks | $\frac{y_{1.}^2}{q} + \cdots + \frac{y_{i.}^2}{q-1} + \cdots + \frac{y_{p.}^2}{q} - \frac{y_{..}'^2}{pq-1}$ | $p-1$ |
| Treatments | (By subtraction) | $q-1$ |
| Error | As in the completed table | $(p-1)(q-1)-1$ |
| Total | $\sum_{i,j} y_{ij}^2 - \frac{y_{00}'^{\,2}}{pq-1}$ | $pq-2$ |

## 5.2 Two-Way Classification with Two Missing Plots

Let the (i,j)-th and (r,s)-th observations be missing. Let $x_1$ and $x_2$ be the values of the missing observations. Then, as in the previous case, the error S.S. is

$$E = x_1^2 + x_2^2 - \frac{(y'_{i.} + x_1)^2}{q} - \frac{(y'_{.j} + x_1)^2}{p}$$

$$- \frac{(y'_{r.} + x_2)^2}{q} - \frac{(y'_{.s} + x_2)^2}{p} + \frac{(y'_{..} + x_1 + x_2)^2}{pq}$$

$$+ \text{ terms independent of } x_1 \text{ and } x_2.$$

Minimizing E w.r.t. $x_1$ and $x_2$, we get the equations

$$x_1 - \frac{y'_{i.} + x_1}{q} - \frac{y'_{.j} + x_1}{p} + \frac{y'_{..} + x_1 + x_2}{pq} = 0 \qquad (1)$$

$$x_2 - \frac{y'_{r.} + x_2}{q} - \frac{y'_{.s} + x_2}{p} + \frac{y'_{..} + x_1 + x_2}{pq} = 0 \qquad (2)$$

Adding (1) and (2) we get

$$x_1 + x_2 = \frac{p(y'_{i.} + y'_{r.}) + q(y'_{.j} + y'_{.s}) - 2y'_{..}}{(p-1)(q-1)+1}.$$

Subtracting (2) from (1) gives

$$x_1 - x_2 = \frac{p(y'_{i.} - y'_{r.}) + q(y'_{.j} - y'_{.s})}{(p-1)(q-1)-1}.$$

These last two equations give $x_1$ and $x_2$.

## 5.3 K-Missing Plots: Yates' Method

In the general case when there are several missing plots, an approximate iterative procedure of Yates is applicable.

As a first approximation all except the first missing plot are filled by the respective block means. Then estimate the value in the first missing plot by the usual formula

$$x = \frac{py'_{i.} + qy'_{.j} - y'_{..}}{(p-1)(q-1)},$$

using the new block and treatment totals. Then, using the new values, a second approximation to the missing value in the second missing plot is obtained. Similarly, second approximations to the remaining missing plots may also be obtained. Usually the second approximations give sufficient accuracy. However, the process may be repeated if more accuracy is needed.

**Completed Table**

| Source | S.S. | d.f. | M.S.S. | F |
|---|---|---|---|---|
| Blocks | $\sum_i \frac{y_{i.}^2}{q} - \frac{y_{..}^2}{pq}$ | $p-1$ | $V_B$ | $F_1$ |
| Treatments | $\sum_j \frac{y_{.j}^2}{p} - \frac{y_{..}^2}{pq}$ | $q-1$ | $V_T$ | $F_2$ |
| Error | (Subtraction) | $(p-1)(q-1)-k$ | $V_E$ | |
| Total | $\sum_{i,j} y_{ij}^2 - \frac{y_{..}^2}{pq}$ | $pq-k-1$ | | |

If $F_2$ is not significant, we may assume that the treatment effects are not significantly different. If $F_2$ is significant the exact test procedure is done using the following analysis of variance table:

| Source | S.S. | d.f. |
|---|---|---|
| Blocks | $\frac{y_{1.}^2}{q}+\cdots+\frac{y_{i.}'^2}{q-1}+\cdots+\frac{y_{r.}'^2}{q-1}+\cdots+\frac{y_{p.}^2}{q}-\frac{y_{..}'^2}{pq-k}$ | $p-1$ |
| Treatments | (By subtraction) | $q-1$ |
| Error | As in the completed table | $(p-1)(q-1)-k$ |
| Total | S. S. of the existing observations $-\frac{y_{..}'^2}{pq-k}$ | $pq-k-1$ |

## 5.4 Bartlett's Method: Use of Concomitant Variables

**Two-way classification with single missing plot**

Let the (i,j)-th observation be missing. Consider a dummy variable x defined by

$$x=\begin{cases}-1, & \text{in the } (i,j)-\text{th cell}\\ 0, & \text{in all other cells}\end{cases}.$$

Consider the following table:

| | 1 | 2 | ... | j | ... | q | Sum |
|---|---|---|---|---|---|---|---|
| 1 | $y_{11}$<br>0 | $y_{12}$<br>0 | ... | $y_{1q}$<br>0 | ... | $y_{1q}$<br>0 | $y_{1.}$<br>0 |
| 2 | $y_{21}$<br>0 | $y_{22}$<br>0 | ... | $y_{2j}$<br>0 | ... | $y_{2q}$<br>0 | $y_{2.}$<br>0 |
| ⋮ | | | ... | | ... | | |
| i | $y_{i1}$<br>0 | $y_{i2}$<br>0 | ... | 0<br>$-1$ | ... | $y_{iq}$<br>0 | $y'_{i.}$<br>$-1$ |
| ⋮ | | | ... | | ... | | |
| p | $y_{p1}$<br>0 | $y_{p2}$<br>0 | ... | $y_{pj}$<br>0 | ... | $y_{pq}$<br>0 | $y_{p.}$<br>0 |
| Sum | $y_{.1}$<br>0 | $y_{.2}$<br>0 | ... | $y'_{.j}$<br>$-1$ | ... | $y_{.q}$<br>0 | $y'_{..}$<br>$-1$ |

Let $Y_{ij} = \alpha_i + \beta_j + bx_{ij} + \varepsilon_{ij}$
$\quad = y_{ij} + bx_{ij}.$

Therefore

$$Y_{ij} = \begin{cases} y_{ij} & \text{when } x_{ij} = 0 \\ -b & \text{when } y_{ij} = 0\,. \end{cases}$$

Hence in (i,j)-th cell $E(Y_{ij}) = -b$, which is to be estimated. The least square estimate of b is $\hat{b} = \frac{E_{xy}}{E_x}$ (with familiar notations). Now

$$E_{xy} = \sum_{i,j} x_{ij}y_{ij} - \sum_{i} \frac{x_{i.}y_{i.}}{q} - \sum_{j} \frac{x_{.j}y_{.j}}{p} + \frac{x_{..}y_{..}}{pq}$$
$$= 0 + \frac{y'_{i.}}{q} + \frac{y'_{.j}}{p} - \frac{y'_{..}}{pq}$$
$$= \frac{py'_{i.} + qy'_{.j} - y'_{..}}{pq}.$$

Similarly, $E_x = \frac{(p-1)(q-1)}{pq}$. Therefore

$$\hat{b} = \frac{py'_{i.} + qy'_{.j} - y'_{..}}{(p-1)(q-1)},$$

which is the same as the estimate obtained by minimizing the error S.S. by Yates method. Here also, $\hat{b}$ is the least square estimate of b. Therefore, the principles also are the same in both cases.

## 5.5 Latin Square

### (i) Single missing plot

Let the (i,j)-th observation corresponding to treatment $T_k$ be missing and let it be denoted by x. The ordinary analysis is done as follows:

| Source | S.S. |
|---|---|
| Rows | $\frac{R_1^2}{s} + \cdots + \frac{(R'_i + x)^2}{s} + \cdots + \frac{R_s^2}{s} - \frac{(G' + x)^2}{s^2}$ |
| Columns | $\frac{C_1^2}{s} + \cdots + \frac{(C'_j + x)^2}{s} + \cdots + \frac{C_s^2}{s} - \frac{(G' + x)^2}{s^2}$ |
| Treatments | $\frac{T_1^2}{s} + \cdots + \frac{(T'_k + x)^2}{s} + \cdots + \frac{T_s^2}{s} - \frac{(G' + x)^2}{s^2}$ |
| Error | (By subtraction) |
| Total | $x^2 - \frac{(G' + x)^2}{s^2}$ + terms not involving x. |

Therefore S.S. is

$$E = x^2 - \frac{(R_i' + x)^2}{s} - \frac{\left(C_j' + x\right)^2}{s} - \frac{(T_k' + x)^2}{s} + \frac{2(G' + x)^2}{s^2}$$
$$+ \text{ terms not involving } x\,.$$

The procedure is to minimize E w.r.t. x. We get

$$x = \frac{s\left(R_i' + C_j' + T_k'\right) - 2G'}{(s-1)(s-2)}.$$

We may replace the missing yield by this estimate and perform the analysis. The d. f. of error S.S. will be $(s-1)(s-2)-1$, and for the total S.S. we have $s^2 - 2$ d. f.. The usual tests of significance can be performed, but the F-ratio will have an upward bias. Thus, if non-significance is found, we can stop. If F is significant, however, we cannot be sure that it is due to treatments and not to this bias.

**Testing individual hypotheses** $\alpha_i = \alpha_l$

When the $(i,j)$-th observation is missing we have

$$\widehat{\alpha}_i - \widehat{\alpha}_l = \frac{R_i' + x}{s} - \frac{R_l}{s} = \frac{1}{s}\left[R_i' + \frac{s\left(R_i' + C_j' + T_k'\right) - 2G'}{(s-1)(s-2)}\right] - \frac{R_l}{s}.$$

$$V\left(\widehat{\alpha}_i - \widehat{\alpha}_j\right) = C\lambda^T\sigma^2 (\text{with } C_1 = 1, C_2 = -1, \lambda_1 = \frac{1}{s} + \frac{1}{(s-1)(s-2)}, \lambda_2 = -\frac{1}{s})$$
$$= \left[\frac{2}{s} + \frac{1}{(s-1)(s-2)}\right]\sigma^2,$$

and for all other contrasts of the form $\alpha_p - \alpha_q$,

$$V\left(\widehat{\alpha}_p - \widehat{\alpha}_q\right) = \frac{2}{s}\sigma^2$$

Hence $\sigma^2$ is best estimated by $V_E$ in the completed analysis of variance table. So we can apply the t-test. The test criterion is

$$t = \frac{\hat{\alpha}_i - \hat{\alpha}_l}{\sqrt{\left[\frac{2}{s} + \frac{1}{(s-1)(s-2)}\right] V_E}} \quad \text{with } (s^2 - 3s + 1) \text{ d.f.}.$$

Similarly, we can test the hypotheses $\beta_j = \beta_l$ and $t_k = t_l$.

**Exact test for treatment effects when the treatment mean square in the completed table is significant**

Let the $(i,j)$-th observation be missing. We have to evaluate the S.S.'s attributable to rows and columns ignoring treatments, and to rows, columns, and treatments. The difference of these two sums of squares may then be tested against the error with the reduced S.S. to complete the necessary quantities. The experiment is first regarded as an experiment in rows and columns with one observation missing and the minimum S.S. for error is obtained as:

$$\text{Row S.S.} = \frac{R_1^2}{s} + \cdots + \frac{R_i'^2}{s-1} + \cdots + \frac{R_s^2}{s} - \frac{G'^2}{s^2-1}$$

$$\text{Column S.S.} = \frac{C_1^2}{s} + \cdots + \frac{C_j'^2}{s-1} + \cdots + \frac{C_s^2}{s} - \frac{G'^2}{s^2-1}$$

$$\text{Total S.S.} = (\text{Sum of squares of known observations}) - \frac{G'^2}{s^2-1}.$$

Then

$$\text{Error S.S.} = \text{Total} - \text{Row S.S.} - \text{Column S.S.} = W, \text{say}.$$

Then W is the error S.S. under the assumption that the treatment effects are the same and will have $(s^2 - 2) - (s - 1) - (s - 1) = (s^2 - 2s)$ d.f..

The minimum S.S. for error, when treatments are taken into account, is obtained by analyzing the augmented table. Let this be E. Then E is the error S.S. when the treatment effects are assumed different and it has $(s^2 - 2) - (s - 1) - (s - 1) - (s - 1) = (s^2 - 3s + 1)$ d.f.. Then $W - E$ will be the treatment S.S. and has $(s - 1)$ d.f.. Hence the test criterion for treatments is

$$F(s-1, s^2-3s+1) = \frac{(W-E)/(s-1)}{E/(s^2-3s+1)}.$$

**(ii) Several Missing Plots**

If there are, say n, missing plots, we proceed exactly as above. However, if n is greater than 2 or 3, the algebra becomes very involved. In such cases, Yates' iterative method will be found more convenient. Usually two iterations will give sufficiently accurate values. More iterations are necessary if greater accuracy is desired.

**Mixed Plots**

Let the observations in the $(i,j)$-th and $(p,q)$-th cells be mixed up, giving a sum A. If x is the $(i,j)$-th observation, then the $(p,q)$-th observation will be $A-x$. Then x is determined by minimizing the error S.S. in the analysis of variance table.

**5.6 Randomized Block (r blocks and v treatments)**

$$\text{Total S.S.} = x^2 + (A-x)^2 + \text{terms independent of x.}$$

$$\text{Row S.S.} = \frac{(y'_{.j}+x)^2 + (y'_{.q}+A-x)^2}{r}$$
$$+ \text{ terms independent of x.}$$

$$\text{Block S.S.} = \frac{(y'_{i.}+x)^2 + (y'_{p.}+A-x)^2}{v}$$
$$+ \text{ terms independent of x.}$$

$$\text{Error S.S.} = E = x^2 + (A-x)^2 - \frac{(y'_{i.}+x)^2 + \left(y'_{p.}+A-x\right)^2}{v}$$
$$-\frac{\left(y'_{.j}+x\right)^2 + \left(y'_{.q}+A-x\right)^2}{r}$$
$$+\text{terms independent of x}.$$

Minimizing E w.r.t. x we get

$$x = \frac{A}{2} + \frac{r(y'_{i.} - y'_{p.})}{2(rv - r - v)} + \frac{v(y'_{.j} - y'_{.q})}{2(rv - r - v)}.$$

**Case when the mixed plots are in the same block**

Let the observations in the (i,j)-th and (i,q)-th cells be mixed up. Then

$$E = x^2 + (A - x)^2 - \frac{(y'_{i\cdot} + A)^2}{v} - \frac{(y'_{\cdot j} + A)^2}{r} - \frac{(y'_{\cdot q} + A - x)^2}{r}$$

$$+\text{terms independent of } x.$$

Minimizing E w.r.t x we get

$$x = \frac{A}{2} + \frac{y'_{.j} - y'_{.q}}{2(r - 1)}.$$

Similarly when the mixed plots are in the same column, say (i,j)-th and (p,j)-th, we get

$$x = \frac{A}{2} + \frac{y'_{i.} - y'_{p.}}{2(v - 1)}.$$

**To test** $\alpha_i = \alpha_l$ (when the (i,j)-th and (p,q)-th cells are mixed up, $l \neq p$).

$$\begin{aligned}\hat{\alpha}_i - \hat{\alpha}_l &= \frac{y_{i\cdot}}{v} - \frac{y_{l\cdot}}{v} = \frac{y'_{i\cdot} + x}{v} - \frac{y_{l\cdot}}{v} \\ &= \frac{1}{v}\left[y'_{i.} + \frac{A}{2} + \frac{r(y'_{i.} - y'_{p.}) + v(y'_{.j} - y'_{.q})}{2(rv - r - v)} - y_{l.}\right]\end{aligned}$$

$$V(\hat{\alpha}_i - \hat{\alpha}_l) = C\lambda^T\sigma^2.$$

with $C_1 = 1, C_2 = -1, \lambda_1 = \frac{1}{v} + \frac{r}{2v(rv - r - v)}, \lambda_2 = -\frac{1}{v}.$

Therefore

$$V(\hat{\alpha}_i - \hat{\alpha}_l) = \left[\frac{2}{v} + \frac{r}{2v(rv - r - v)}\right]\sigma^2,$$

and $\sigma^2$ is best estimated by $V_E$, the error mean square in the analysis of variance table. So we can apply the t-test. Similarly we can test for $\beta_j = \beta_l$ $(l \neq q)$.

**Latin Square**

In this case, from the ordinary analysis, we get

$$E = x^2 + (A - x)^2 - \frac{(R_i' + x)^2}{s} - \frac{(R_p' + A - x)^2}{s} - \frac{(C_j' + x)^2}{s} - \frac{(C_q' + A - x)^2}{s} - \frac{(T_k' + x)^2}{s} - \frac{(T_r' + A - x)^2}{s}$$
$$+ \text{ terms independent of } x.$$

Here the mixed up observations are the (i,j)-th and (p,q)-th. The (i,j)-th observation corresponds to treatment $T_r$.
Minimizing E w.r.t. x we get

$$x = \frac{A}{2} + \frac{(R_i' - R_p') + (C_j' - C_q') + (T_k' - T_r')}{2(s-3)}.$$

**Case when the mixed up observations correspond to the same treatment, say $T_k$**

$$E = x^2 + (A - x)^2 - \frac{(R_i' + x)^2}{s} - \frac{(R_p' + A - x)^2}{s} - \frac{(C_j' + x)^2}{s} - \frac{(C_q' + A - x)^2}{s} - \frac{(T_k' + A)^2}{s}$$
$$+ \text{ terms independent of } x.$$

Here

$$x = \frac{A}{2} + \frac{(R_i' - R_p') + (C_j' - C_q')}{2(s-2)}.$$

**Case when the mixed up observations are in the same row**

Le the (i,j)-th and (i,q)-th observations be mixed up. Let the corresponding observations be $T_k$ and $T_r$. Then

$$E = x^2 + (A-x)^2 - \frac{(C_j' + x)^2}{s} - \frac{(C_q' + A - x)^2}{s} - \frac{(T_k' + x)^2}{s} - \frac{(T_r' + A - x)^2}{s}$$

$$+ \text{ terms independent of } x.$$

Minimizing E w.r.t. x we get

$$x = \frac{A}{2} + \frac{(C_j' - C_q') + (T_k' - T_r')}{2(s-2)}.$$

Similarly, if the mixed up observations are in the same column, say the (i,j)-th and (p,j)-th observations are mixed up, then we have

$$x = \frac{A}{2} + \frac{(R_i' - R_p') + (T_k' - T_r')}{2(s-2)}.$$

**General Case – Iterative method of Yates**

Let there be k mixed up cells, say in a randomized block experiment, with sum A. Substitute the value $\frac{A}{k}$ in these cells. Then $\frac{A}{k}$ will give a first approximation to the mixed up observations. A second approximation is obtained as follows.

Consider the first two mixed up cells. Let the observation in the first cell be x. Then the observation in the second cell is

$$y = A - x - \left(\frac{k-2}{k}\right)A$$
$$= \frac{2A}{k} - x.$$

Considering the observations in the other $(k-2)$ mixed up cells as $\frac{A}{k}$, minimize the present error S.S. with respect to x. This will give a second approximation to the observations in the first two cells. By a similar procedure we can get second approximation to the observations in the other mixed up cells. The procedure can be repeated.

**Bartlett's method of Concomitant variables**

Consider a randomized block with two mixed up cells. Let the (i,j)-th and (p,q)-th cells be mixed up. Consider a concomitant variable x defined by

$$x = \begin{cases} +1 & \text{in the } (i,j)-\text{th cell} \\ -1 & \text{in the } (p,q)-\text{th cell} \\ 0 & \text{in all other cells.} \end{cases}$$

Let the regression model be

$$Y_{ij} = y_{ij} + bx_{ij}.$$

Then $Y_{ij} = \frac{A}{2} + b$ in the $(i,j)-$th cell

$= \frac{A}{2} - b$ in the $(p,q)-$th cell.

We have $\hat{b} = \frac{E_{xy}}{E_x}$,

$$E_{xy} = \sum_{i,j} x_{ij}y_{ij} - \sum_i \frac{x_{i.}y_{i.}}{v} - \sum_j \frac{x_{.j}y_{.j}}{r} + \frac{x_{..}y_{..}}{rv}$$
$$= 0 - \frac{\left(y'_{i.} + \frac{A}{2}\right) - \left(y'_{p.} + \frac{A}{2}\right)}{v} - \frac{\left(y'_{.j} + \frac{A}{2}\right) - \left(y'_{.q} + \frac{A}{2}\right)}{r}$$
$$= -\left[\frac{y'_{i.} - y'_{p.}}{v} - \frac{y'_{.j} - y'_{.q}}{r}\right],$$

and

$$E_x = \sum_{i,j} x_{ij}^2 - \sum_i \frac{x_{i.}^2}{v} - \sum_j \frac{x_{.j}^2}{r} + \frac{x_{..}^2}{rv}$$
$$= \frac{2}{rv}(rv - r - v)\,.$$

The mixed up values are

$$x = \frac{A}{2} + \frac{v(y'_{.j} - y'_{.q}) - r(y'_{i.} - y'_{p.})}{2(rv - r - v)} \text{ and } Y = A - x.$$

**Analysis**

When there are 2 mixed up plots, one $\mathrm{d.f.}$ will be reduced from the total and hence from the error $\mathrm{d.f.}$ also. When there are k mixed up plots, $(k-1)$ $\mathrm{d.f.}$ will be reduced.

After the missing values or the mixed values are estimated the final ANOVA table is written. Since the ANOVA table includes only the squares of these estimated missing values, the positive or the negative signs of the missing estimated values do not matter in the final ANOVA table. This is because in normal distribution the observed value of the random variable or variate or random vector may be positive or negative but the random vector has the same distribution. The estimator has the same distribution whether estimate is positive or negative.

# Chapter 6

# Balanced Incomplete Block Designs

## 6.1 Introduction

When the number of treatments to be tried is very large, it will be very difficult to get homogeneous blocks large enough to accommodate all the treatments. There are two main groups of designs in this case:

(i) Quasi factorial or Lattice Designs, in which a correspondence maybe established between the treatments and the treatment combinations of a factorial set. The method of analysis in this case is the same as that for the factorial designs when the treatments are renamed according to the above correspondence.

(ii) Incomplete Blocks in which, although it may be possible to set up the sort of correspondence with the factorial set up, it is of little or no help in devising the design or simplifying the analysis.

The property of balance in an incomplete block design is that each possible pair of treatments occurs together in the same number (denoted by $\lambda$) of blocks. Thus a system of v treatments arranged in b blocks of k ($<$v) plots each such that no treatment occurs twice in the same block is said to form a Balanced Incomplete Block Design (BIBD) if each treatment occurs in exactly r blocks and each pair of treatments occurs together in $\lambda$ blocks.

It is readily seen that $bk = vr$, each side denoting the total number of plots used. A particular variety can be paired with all the $(v - 1)$ remaining varieties each of which is occurring in $\lambda$ blocks. But that particular variety is occurring in r blocks in each of which it can form $(k - 1)$ pairs with the remaining varieties occurring in the $(k - 1)$ plots. Thus we get

$$\lambda(v - 1) = r(k - 1).$$

Further $\lambda$ must be less than r for if $\lambda = r$ it means that in all the blocks in which a particular variety occurs, all the remaining varieties also should occur, but then the design becomes a randomized block design (k=v). So $r > \lambda$. [This can be directly seen from the relation $\lambda(v-1) = r(k-1)$ for $v > k$ in the design.]

**Theorem 6.1 (Fisher's Inequality):**

$$b \geq v.$$

**Proof.**

Let $n_{ij} = 1$, if the j-th variety occurs in the i-th block
$= 0$, otherwise.

Then clearly,

$$\sum_i n_{ij} n_{il} = \lambda, \text{if } j \neq l,$$

since the pair $(j, l)$ occurs together in $\lambda$ blocks. If $j = l$, it becomes

$$\sum_i n_{ij}^2 = r,$$

since the treatment j occurs in r blocks.

If possible, let $b < v$. Now consider the $v \times v$ matrix

$$N = \begin{bmatrix} n_{11} & n_{21} & \dots & n_{b1} & 0 & \dots & 0 \\ n_{12} & n_{22} & \dots & n_{b2} & 0 & \dots & 0 \\ \dots & \dots & \dots & \dots & \dots & \dots & \dots \\ n_{1v} & n_{2v} & \dots & n_{bv} & 0 & \dots & 0 \end{bmatrix}$$

where the last $v - b$ columns of N are zero vectors.

From the relations among the $n_{ij}$'s, we get

$$NN^T = \begin{bmatrix} r & \lambda & \dots & \lambda \\ \lambda & r & \dots & \lambda \\ \dots & \dots & \dots & \dots \\ \lambda & \lambda & \dots & r \end{bmatrix}.$$

Therefore

$$\begin{aligned} |NN^T| &= [r + \lambda(v-1)](r-\lambda)^{v-1} \\ &= [r + r(k-1)](r-\lambda)^{v-1} \\ &= rk(r-\lambda)^{v-1}. \end{aligned}$$

But from the definition of N, $|N| = 0$. Therefore

$$|NN^T| = |N||N^T| = 0.$$

$$\text{i.e.}\quad rk(r-\lambda)^{v-1} = 0,$$

which implies $r = \lambda$. This is contradictory. Therefore $b \geq v$.

Since $bk = vr$, we get

$$\frac{b}{v} = \frac{r}{k} \text{ or } \frac{b-v}{v} = \frac{r-k}{k}.$$

$$\text{i.e.,}\, b - v = \frac{v}{k}(r-k)$$

But we have proved that $b - v \geq 0$, and $\frac{v}{k} > 1$. Therefore

$$b - v \geq r - k$$

$$\text{i.e.,}\, b \geq v + r - k.$$

Again,

$$v > k \quad \text{i.e.,}\, v \geq k + 1$$

and hence

$$b \geq v \geq k + 1. \blacksquare$$

## 6.2 Symmetrical Balanced Incomplete Blocks

If $b = v$ (so that $r = k$), the design is said to be symmetrical. In this case the above inequalities can be modified for $v > k$.
Therefore

$$b = v \geq k + 1.$$

Again,

$$\lambda(v-1) = r(k-1) = k(k-1)$$

$$\text{or } \lambda = \frac{k}{v-1}(k-1).$$

But since

$$v \geq k+1, \quad \frac{k}{v-1} \leq 1,$$

we have

$$\lambda \leq k-1 \quad \text{or} \quad \lambda + 2 \leq k+1.$$

In any BIBD, $b \geq v \geq k+1$. Therefore in a symmetric BIBD,

$$b = v \geq k+1 \geq \lambda + 2.$$

**Theorem 6.2**

In a symmetrical BIBD, any two blocks have exactly $\lambda$ varieties in common, a property which is useful in the construction of BIBD's.

**Proof.**

Let $x_{ij}$ denote the number of treatments common to the i-th and j-th blocks. Then for $i \neq j$,

$$\sum_i x_{ij} = k(r-1),$$

because altogether there are k varieties in the j-th block and each variety can occur in $(r-1)$ other blocks. But in a symmetrical BIBD, $r = k$. Therefore

$$\sum_i x_{ij} = r(k-1)$$

$$= \lambda(v-1), i \neq j.$$

Now, the pairs that are common to the i-th and j-th blocks is $\frac{x_{ij}(x_{ij}-1)}{2}$, i.e., $\binom{x_{ij}}{2}$. Therefore considering the possible pairs that can be obtained with the k varieties that occur in the i-th block, with those of the j-th block, where $i \neq j$, we get

$$\sum_{i} \frac{x_{ij}(x_{ij} - 1)}{2} = (\lambda - 1)\frac{k(k - 1)}{2},$$

because each pair in the j-th block occurs in $(\lambda - 1)$ of the remaining blocks. So,

$$\begin{aligned}\sum_{i} x_{ij}^2 &= \sum_{i} x_{ij} + (\lambda - 1)k(k - 1) \\ &= k(k - 1) + (\lambda - 1)k(k - 1) \\ &= \lambda k(k - 1) \\ &= \lambda r(k - 1) \\ &= \lambda^2(v - 1).\end{aligned}$$

Further,

$$\begin{aligned}\sum_{i}(x_{ij} - \lambda)^2 &= \sum_{i}(x_{ij}^2 - 2\lambda x_{ij} + \lambda^2) \\ &= \sum_{i} x_{ij}^2 - 2\lambda \sum_{i} x_{ij} + \lambda^2 \sum_{i}(1) \\ &= \sum_{i} x_{ij}^2 - 2\lambda r(k - 1) + (v - 1)\lambda^2,\end{aligned}$$

since, for $i \neq j$, $\sum_i(1) = b - 1 = v - 1$, and

$$\begin{aligned}\sum_{i}(x_{ij} - \lambda)^2 &= \sum_{i} x_{ij}^2 - 2\lambda^2(v - 1) + \lambda^2(v - 1) \\ &= \sum_{i} x_{ij}^2 - \lambda^2(v - 1) \\ &= 0,\end{aligned}$$

which implies $x_{ij} - \lambda = 0$ for each $j \neq i$. Therefore $x_{ij} = \lambda$. Hence the number of varieties common to any two blocks is the same, viz. $\lambda$. ■

## 6.3 Resolvable Balanced Incomplete Blocks

If it is possible to separate the b blocks into r sets of n blocks each $(b = nr)$ so that each variety occurs once and only once among the blocks of a given set, i.e., each set forms a complete replicate, then the design is said to be resolvable. Since, each set forms a complete replication, the total number of plots in the set is $nk = v$. For a resolvable balanced incomplete block, the inequality $b \geq v$ can be replaced by a more stringent inequality given below,

**Theorem 6.3**

For a resolvable BIBD $b \geq v + r - 1$.

**Proof.**

Let $l_{ij}$ denote the number of varieties common to the block $B_{11}$ (i.e., 1st block in the 1st set) and the block $B_{ij}$ (i.e., j-th block in the i-th set), $i \neq 1$. Then the $n(r-1)$ number $l_{ij}$ sum to $k(r-1)$, for, each of the k varieties in $B_{11}$ will be occurring in $(r-1)$ of the blocks of the remaining sets $(i \neq 1)$. Similarly, by taking into account the possible number of pairs (distinct) among these blocks, we get,

$$\sum_{i,j} \frac{l_{ij}(l_{ij}-1)}{2} = (\lambda - 1)\frac{k(k-1)}{2}, \quad i \neq 1.$$

Therefore

$$\begin{aligned}
\sum_{i,j} l_{ij}^2 &= \sum_{i,j} l_{ij} + (\lambda - 1)k(k-1) \\
&= k(r-1) + (\lambda - 1)k(k-1) \\
&= k[r - 1 + (\lambda - 1)(k-1)] \\
&= k[r - k + \lambda(k-1)] \\
&= k\left[\frac{(r-k)(nk-1) + r(k-1)^2}{(nk-1)}\right], \quad \text{since } \lambda = \frac{r(k-1)}{nk-1}.
\end{aligned}$$

If we denote by $S^2$ the variance of the $b_{ij}$'s,

$$S^2 = \frac{\sum l_{ij}^2}{n(r-1)} - [\frac{k(r-1)}{n(r-1)}]^2$$

$$= \frac{k}{n(nk-1)(r-1)}[(r-k)(nk-1) + r(k-1)^2] - \frac{k^2}{n^2}$$

$$= \frac{k}{n^2(nk-1)(r-1)}[n(nkr - r - nk^2 + k) + nr(k^2 - 2k + 1) - k(nkr - nk - r + 1)]$$

$$= \frac{k^2}{n^2(nk-1)(r-1)}[n(nr - nk + 1) - 2nr + nk + r - 1]$$

$$= \frac{k^2}{n^2(nk-1)(r-1)}[r(n^2 - 2n + 1) - (nk - 1)(n - 1)]$$

$$= \frac{(n-1)k^2}{n^2(nk-1)(r-1)}[r(n-1) - (nk-1)]$$

$$= \frac{k(nk-k)}{n^2(nk-1)(r-1)}[nr - r - nk + 1]$$

$$= \frac{k(v-k)}{n^2(nk-1)(r-1)}[b - r - v + 1] \geq 0.$$

But $v > k$ and hence $b - r - v + 1 \geq 0$. Therefore $b \geq v + r - 1$. ■

We have also seen that the mean of the $l_{ij}$'s is

$$m = E(l_{ij}) = \frac{k(r-1)}{n(r-1)} = \frac{k}{n} = \frac{k^2}{v}.$$

## 6.4 Affine Resolvable Balanced Incomplete Blocks

If a resolvable balanced incomplete block is such that any two blocks belonging to two different sets have always the same number of varieties in common, then the design is said to be affine resolvable. Since in an affine resolvable design the $b'_{ij}$'s are to be equal, their variance is zero, which is possible only when $b = v + r - 1$. So the necessary and sufficient condition that a resolvable balanced incomplete block is affine resolvable is that $b = v + r - 1$. It should be noted that all designs in which $b = v + r - 1$ are not affine resolvable. Since each $b_{ij} = m = \frac{k^2}{v}$ (proved earlier) we get $k^2$ must be an integral multiple of v, which can be used to verify whether a given balanced incomplete block satisfying the condition $b = v + r - 1$ is affine resolvable or not. Actually $b = v + r - 1$ is only a necessary condition for a balanced incomplete block to be affine resolvable.

### Efficiency of BIBD's

The variance of any two treatment comparisons in this design is $\frac{2k}{\lambda v}\sigma^2$ where $\sigma^2$ is estimated by the error variance. If we had used a randomized block having the same error variance, the efficiency of the BIBD compared to the randomized block is

$$\frac{\frac{2V_E}{r}}{\frac{2k}{\lambda v}V_E} = \frac{\lambda v}{rk} = \frac{rv(k-1)}{rk(v-1)} = \frac{1-\frac{1}{k}}{1-\frac{1}{v}}.$$

### Theorem 6.4 (An existence theorem for BIBD's)

A symmetric BIBD with an even number of varieties exists if $(r - \lambda)$ is a perfect square.

**Proof.**

Consider the $v \times v$ matrix

$$N = \begin{bmatrix} n_{11} & n_{21} & \dots & n_{v1} \\ n_{12} & n_{22} & \dots & n_{v2} \\ \dots & \dots & \dots & \dots \\ n_{1v} & n_{2v} & \dots & n_{vv} \end{bmatrix},$$

where

$$n_{ij} = 1, \text{if the j-th variety occurs in the i-th block}$$
$$= 0, \text{otherwise.}$$

Then $|NN^T| = rk(r-\lambda)^{v-1}$

$$= k^2(r-\lambda)^{v-1}, \quad \text{since } r = k$$

i.e., $|N|^2 = k^2(r-\lambda)^{v-1}$.

Note that $|N|^2$ is a perfect square and $k^2$ is also a perfect square. When v is even, v-1 is odd. Hence $(r-\lambda)^{v-1}$ is a perfect square only if $r-\lambda$ itself is a perfect square. ■

## 6.5 Complementary BIBD

Consider the following BIBD with parameters

$$v = 7, \ b = 7, \ k = 3, \ r = 3, \ \lambda = 1.$$

| | | |
|---|---|---|
| 1 | 2 | 4 |
| 2 | 3 | 5 |
| 3 | 4 | 6 |
| 4 | 5 | 7 |
| 5 | 6 | 1 |
| 6 | 7 | 2 |
| 7 | 1 | 3 |

The complementary BIBD is the design with blocks containing treatments that do not occur in the corresponding blocks of a BIBD. The complementary BIBD of the above BIBD is

$$b' = 7, \ v' = 7, \ k' = 4, \ r' = 4, \ \lambda' = 2$$

| 3 | 5 | 6 | 7 |
|---|---|---|---|
| 1 | 4 | 6 | 7 |
| 1 | 2 | 5 | 7 |
| 1 | 2 | 3 | 6 |
| 2 | 3 | 4 | 7 |
| 1 | 3 | 4 | 5 |
| 2 | 4 | 5 | 6 |

The complementary BIBD is also a BIBD.

**Example 6.1. Resolvable BIBD**

$$\left.\begin{array}{cc} 1 & 2 \\ 3 & 4 \end{array}\right\} \text{1-st replicate}$$
$$\left.\begin{array}{cc} 1 & 3 \\ 2 & 4 \end{array}\right\} \text{2-nd replicate}$$
$$\left.\begin{array}{cc} 1 & 4 \\ 2 & 3 \end{array}\right\} \text{3-rd replicate}$$

## 6.6 Analysis of BIBD Experiment

As in the case of quasi factorial designs there are two types of information on treatment comparisons provided by the experiment: intra-block derived from comparisons within blocks, and inter-block from comparisons between blocks.

### (i) Intra-Block Analysis

Consider a BIBD with parameters $v, b, r, k, \lambda$. Let $y_{ij}$ denote the observation in the $i$-th block corresponding to the $j$-th variety (if it occurs in that block, of course). Let $\alpha_i$ be the effect of the $i$-th block and $\beta_j$ the effect of the $j$-th variety. Let

$$n_{ij} = 1, \text{if the j-th variety is in the i-th blocks,}$$
$$= 0, \text{otherwise.}$$

Then, assuming additivity, the model for the intrablock comparison is

$$y_{ij} = n_{ij}(\alpha_i + \beta_j) + \varepsilon_{ij},$$

where $\varepsilon_{ij}$'s are uncorrelated random errors which we assume to be normally distributed around zero with a common variance $\sigma^2$. Therefore the observational equations are

$$E(y_{ij}) = n_{ij}(\alpha_i + \beta_j),\ \ V(y_{ij}) = \sigma^2,\ \ i = 1,2,\dots,b;\ \ j = 1,2,\dots,v$$

Now,

$$\sum_i n_{ij} = n_{.j} = r,$$

for each j, since each variety occurs in r blocks. Similarly,

$$\sum_i n_{ij}^2 = r,\ \ \text{since}\ \ n_{ij}^2 = n_{ij}.$$

Again, for j and l,

$$\sum_i n_{ij}n_{il} = \lambda,\ \ \text{for every j and l},$$

since a pair of treatments occurs in $\lambda$ blocks.

Also, $\sum_j n_{ij} = k$, since there are k varieties in one block, and $\sum_j n_{ij}^2 = k$. The total error S.S. is

$$L = \sum_{i,j} [y_{ij} - n_{ij}(\alpha_i + \beta_j)]^2.$$

Minimizing L w.r.t. $\alpha_i$ and $\beta_j$, the normal equations are

$$\sum_j n_{ij}(y_{ij} - n_{ij}\alpha_i - n_{ij}\beta_j) = 0,\ \ i = 1,\dots,b$$

$$\text{and}\ \ \sum_i n_{ij}(y_{ij} - n_{ij}\alpha_i - n_{ij}\beta_j) = 0,\ \ j = 1,\dots,v$$

$$\text{i.e.,}\ \ \sum_j n_{ij}y_{ij} - \sum_j n_{ij}^2\alpha_i - \sum_j n_{ij}^2\beta_j = 0,$$

$$\text{and} \quad \sum_i n_{ij}y_{ij} - \sum_i n_{ij}^2\alpha_i - \sum_i n_{ij}^2\beta_j = 0,$$

$$\text{i.e.,} \quad \sum_j n_{ij}y_{ij} - k\alpha_i - \sum_j n_{ij}\beta_j = 0,$$

$$\text{and} \quad \sum_i n_{ij}y_{ij} - \sum_i n_{ij}\alpha_i - r\beta_j = 0.$$

Now, $\sum_j n_{ij}y_{ij}$ is the i-th block total. Let us denote it by $B_i$. Similarly, $\sum_i n_{ij}y_{ij}$ is the sum of the observations corresponding to the j-th variety. Let it be denoted by $V_j$. Then the normal equations take the form

$$B_i - k\alpha_i - \sum_j n_{ij}\beta_j = 0 \tag{1}$$

$$\text{and} \quad V_j - \sum_i n_{ij}\alpha_i - r\beta_j = 0 \tag{2}$$

Altogether there are $b + v$ equations. But $\sum_i(1) = \sum_j(2)$, so only $(b + v - 1)$ equations are independent. So we can impose one subsidiary condition for solving them and this we shall do later. Form (1),

$$\hat{\alpha}_i = \frac{B_i}{k} - \frac{1}{k}\sum_j n_{ij}\beta_j = \frac{B_i}{k} - \frac{1}{k}\sum_l n_{il}\beta_l.$$

Substituting in (2),

$$V_j - \sum_i n_{ij}\left(\frac{B_i}{k} - \frac{1}{k}\sum_l n_{il}\beta_l\right) - r\beta_j = 0.$$

$$\text{i.e.,} \quad V_j - \frac{1}{k}\sum_i n_{ij}B_i + \frac{1}{k}\sum_i n_{ij}\sum_l n_{il}\beta_l - r\beta_j = 0.$$

$$\text{i.e.,} \quad V_j - \frac{1}{k}\sum_i n_{ij}B_i + \frac{1}{k}\sum_l \beta_l \sum_i n_{il}n_{ij} - r\beta_j = 0.$$

Now,

$$\sum_l \beta_l \sum_i n_{il} n_{ij}$$
$$= \beta_1 \sum_i n_{i1} n_{ij} + \beta_2 \sum_i n_{i2} n_{ij} + \cdots$$
$$+ \beta_{j-1} \sum_i n_{i,j-1} n_{ij}$$

$$+\beta_j \sum_i n_{ij}^2 + \beta_{j+1} \sum_i n_{i,j+1} n_{ij} + \cdots + \beta_v \sum_i n_{iv} n_{ij}$$
$$= \beta_1 \lambda + \cdots + \beta_{j-1}\lambda + \beta_j r + \beta_{j+1}\lambda + \cdots + \beta_v \lambda$$
$$= \lambda(\beta_1 + \cdots + \beta_{j-1} + \beta_j + \beta_{j+1} + \cdots + \beta_v) + (r - \lambda)\beta_j$$
$$= \lambda \beta_. + (r - \lambda)\beta_j$$

Therefore

$$V_j - \frac{1}{k} \sum_i n_{ij} B_i + \frac{1}{k}[\lambda \beta_. + (r - \lambda)\beta_j] - r\beta_j = 0.$$

$$\text{i.e.,} \quad V_j - \frac{1}{k} \sum_i n_{ij} B_i + \frac{\lambda}{k}\beta_. + \left(\frac{r - \lambda}{k} - r\right)\beta_j = 0.$$

Now,

$$\frac{r - \lambda}{k} - r = -\frac{r(k-1) + \lambda}{k} = -\frac{\lambda(v-1) + \lambda}{k} = -\frac{\lambda v}{k}.$$

Hence

$$V_j - \frac{1}{k} \sum_i n_{ij} B_i + \frac{\lambda}{k}\beta_. - \frac{\lambda v}{k}\beta_j = 0.$$

But $\sum_i n_{ij} B_i$ is the sum of those block totals where the j-th variety occurs. Let us denote it by $T_j$. Then we have

$$V_j - \frac{1}{k}T_j + \frac{\lambda}{k}\beta_. - \frac{\lambda v}{k}\beta_j = 0.$$

Imposing the condition $\beta_. = 0$, we get

$$\widehat{\beta}_j = \frac{k}{\lambda v}V_j - \frac{1}{\lambda v}T_j = Q'_j, \text{ say.}$$

It is easy to see that $\sum_j Q'_j = 0$, in conformity with the condition we have imposed. Substituting $\widehat{\beta}_j$ in $\widehat{\alpha}_i$, we get

$$\widehat{\alpha}_i = \frac{B_i}{k} - \frac{1}{k}\sum_l n_{il}\widehat{\beta}_l$$
$$= \frac{B_i}{k} - \frac{1}{k}\sum_l n_{il}\left(\frac{k}{\lambda v}V_l - \frac{1}{\lambda v}T_j\right)$$
$$= \frac{B_i}{k} - \frac{1}{\lambda v}\sum_l n_{il}V_l + \frac{1}{k\lambda v}\sum_l n_{il}T_l.$$

Therefore the unconditional minimum S.S. is

$$S^2 = \sum_{i,j} y_{ij}(y_{ij} - n_{ij}\widehat{\alpha}_i - n_{ij}\widehat{\beta}_j)$$
$$= \sum_{i,j} n_{ij}y_{ij}(y_{ij} - \widehat{\alpha}_i - \widehat{\beta}_j), \text{ by definition of } n_{ij}\text{'s}$$
$$= \sum_{i,j} n_{ij}y_{ij}^2 - \sum_i \widehat{\alpha}_i \sum_j n_{ij}y_{ij} - \sum_j \widehat{\beta}_j \sum_i n_{ij}y_{ij}$$
$$= \sum_{i,j} n_{ij}y_{ij}^2 - \sum_i \widehat{\alpha}_i B_i - \sum_j \widehat{\beta}_j V_j$$
$$= \sum_{i,j} n_{ij}y_{ij}^2 - \sum_i \widehat{\alpha}_i B_i - \sum_j Q'_j V_j,$$

with $vr - (b + v - 1)$ or $bk - (b + v - 1)$ d. f. .

Let us consider the hypothesis,

$$H_0: \ \beta_1 = \beta_2 = \cdots = \beta_v (= \beta, \ \text{say}).$$

The total error S.S. under $H_0$ is

$$L_0 = \sum_{i,j} n_{ij}(y_{ij} - \alpha_i - \beta)^2,$$

and the least square normal equations are

$$\sum_{j} n_{ij}(y_{ij} - \alpha_i - \beta) = 0,$$

and

$$\sum_{i,j} n_{ij}(y_{ij} - \alpha_i - \beta) = 0.$$

The first equation gives

$$\sum_{j} n_{ij}y_{ij} - (\alpha_i + \beta)\sum_{j} n_{ij} = 0.$$

$$\text{i.e.,} \ \ B_i - k(\alpha_i + \beta) = 0.$$

Thus the joint estimate of $\alpha_i + \beta$ is

$$\widehat{(\alpha_i + \beta)} = \frac{B_i}{k}.$$

So the conditional minimum S.S. is

$$\begin{aligned} S_0^2 &= \sum_{i,j} n_{ij}(y_{ij} - \frac{B_i}{k})y_{ij} \\ &= \sum_{i,j} n_{ij}y_{ij}^2 - \frac{1}{k}\sum_{i} B_i \sum_{j} n_{ij}y_{ij} \\ &= \sum_{i,j} n_{ij}y_{ij}^2 - \frac{1}{k}\sum_{i} B_i^2. \end{aligned}$$

Hence the S.S. due to treatments (where the S.S. due to blocks is already eliminated) is

$$S_0^2 - S^2 = \sum_i \hat{\alpha}_i B_i + \sum_j V_j Q'_j - \sum_i \frac{B_i^2}{k},$$

with d.f. $(v - 1)$.
Now,

$$\begin{aligned}
S_0^2 - S^2 &= \sum_i \hat{\alpha}_i B_i + \sum_j V_j Q'_j - \sum_i \frac{B_i^2}{k} \\
&= \sum_i \Big(\frac{B_i}{k} - \frac{1}{k}\sum_j n_{ij}\hat{\beta}_j\Big) B_i + \sum_j V_j Q'_j - \sum_i \frac{B_i^2}{k} \\
&= \sum_j V_j Q'_j - \frac{1}{k}\sum_j \hat{\beta}_j \sum_i n_{ij} B_i \\
&= \sum_j V_j Q'_j - \frac{1}{k}\sum_i Q'_j T_j \\
&= \sum_j \Big(V_j - \frac{1}{k} T_j\Big) Q'_j \\
&= \sum_j \frac{\lambda v}{k} Q_j'^2 \\
&= \sum_j \frac{\lambda v}{k}\left(\frac{kQ_j}{\lambda v}\right)^2, \quad \text{if } Q_j = \frac{\lambda v}{k} \quad Q'_j = \frac{\lambda v}{k}\left(\frac{k}{\lambda v} V_j - \frac{1}{\lambda v} T_j\right) \\
&\qquad = V_j - \frac{1}{k} T_j \\
&= \frac{1}{\lambda v k}\sum_j (kQ_j)^2.
\end{aligned}$$

Therefore S.S. due to treatments (eliminating blocks) is

$$\frac{1}{\lambda v k}\sum_{j}(kQ_j)^2$$

with $(v-1)$ d.f., where $Q_j = V_j - \frac{1}{k}T_j$.

**(ii) Intra-Block Analysis of Variance**

| Source | S.S. | d.f. | M.S.S | F |
|---|---|---|---|---|
| Blocks (ignoring treatments) | $\sum_{i}\frac{B_i^2}{k}-\frac{G^2}{bk}$ | $b-1$ | | |
| Treatments (eliminating blocks) | $\frac{1}{\lambda v k}\sum_{j}(kQ_j)^2$ | $v-1$ | $V_T$ | $V_T/V_E$ |
| Error (intra-block) | (By sub) | (sub) | $V_E$ | |
| Total | $\sum_{i,j}y_{ij}^2-\frac{G^2}{bk}$ | $bk-1$ | | |

(G = grand total)

Here we can test only the treatment effects. The information, $I = 1/V_E$, from this table is called the intra-block information.

Since the effects are orthogonal, the S.S. satisfies the equality relation

Blocks (ignoring treatments) + Treatments (eliminating blocks)
= Blocks (eliminating treatments)
+ treatments (ignoring blocks) = Total − Error.

(Here Error means intra-block error). Hence for testing block effects the analysis of variance table will be as follows:

| Source | S. S. | d. f. | M. S. S | F |
|---|---|---|---|---|
| Blocks (eliminating treatment) | (By sub) | $b-1$ | $V_B$ | $V_B$ $/V_E$ |
| Treatments (ignoring treamants) | $\sum_i \frac{V_j^2}{r} - \frac{G^2}{bk}$ | $v-1$ | | |
| Error (intra-block) | (From intra-block analysis) | (sub) | $V_E$ | |
| Total | $\sum_{i,j} y_{ij}^2 - \frac{G^2}{bk}$ | $bk-1$ | | |

**(iii) Inter-Block Information**

Consider a randomized block design with v treatments and r blocks. Let $\alpha_i$ be the i-th block effect and $\beta_j$ the j-th treatment effect. Then we set up the additive model

$$y_{ij} = \alpha_i + \beta_j + \varepsilon_{ij},$$

where $\varepsilon$'s are random errors which are independently and normally distributed around zero with a common variance $\sigma^2$. It is further assumed that the block effects are uncorrelated with the treatment effects and also with the $\varepsilon_{ij}$'s. With this model the p-th and q-th block sums are

$$B_p = v\alpha_p + \beta_. + \sum_{j=1}^{v} \varepsilon_{pj}$$

$$\text{and } B_q = v\alpha_q + \beta_. + \sum_{j=1}^{v} \varepsilon_{qj}.$$

Therefore

$$\frac{B_p}{v} - \frac{B_q}{v} = (\alpha_p - \alpha_q) + (\bar{e}_p - \bar{e}_q),$$

where

$$\bar{e}_p = \frac{1}{v}\sum_{j=1}^{v} \varepsilon_{pj}.$$

$$\text{i.e.,}\ \ \bar{B}_p - \bar{B}_q = (\alpha_p - \alpha_q) + (\bar{e}_p - \bar{e}_q).$$

Under the assumption made above, we see that $(\bar{B}_p - \bar{B}_q)$ is distributed normally around the mean $(\alpha_p - \alpha_q)$ with a variance $\frac{2\sigma^2}{v}$. Here all the treatment effects cancel and so we get no information regarding the treatments from such a block comparison. This is the case with all designs which are complete w.r.t. blocks.

Now consider a BIBD with parameters $b, v, k, r, \lambda$. The $p$-th and $q$-th block sums are

$$B_p = k\alpha_q + \sum_{j=1}^{v} n_{pj}\beta_j + \sum_{j=1}^{k} \varepsilon_{pj}$$

$$\text{and}\ \ B_q = k\alpha_q + \sum_{j=1}^{v} n_{qj}\beta_j + \sum_{j=1}^{k} \varepsilon_{qj}.$$

Therefore

$$\bar{B}_p = \alpha_p + \frac{1}{k}\sum_{j=1}^{v} n_{pj}\beta_j + \bar{e}_p$$

$$\text{and}\ \ \bar{B}_q = \alpha_q + \frac{1}{k}\sum_{j=1}^{v} n_{qj}\beta_j + \bar{e}_q\,.$$

Therefore

$$\overline{B}_p - \overline{B}_q = (\alpha_p - \alpha_q) + \frac{1}{k}\sum_{j=1}^{v}(n_{pj} - n_{qj})\beta_j + (\bar{e}_p - \bar{e}_q).$$

Thus the comparison between the two block means includes a contrast of the p-th and q-th block effects, an expression in terms of some treatment effects and an error term. [The term $\sum_{j=1}^{v}(n_{pj} - n_{qj})\beta_j$ does not vanish since all the treatments that occur in the p-th block do not occur in the q-th.] Under the assumptions made earlier, $\overline{B}_p - \overline{B}_q$ is distributed normally around the mean $(\alpha_p - \alpha_q) + \frac{1}{k}\sum_{j=1}^{v}(n_{pj} - n_{qj})\beta_j$ and variance $\frac{2\sigma^2}{v}$. Since all the treatment effects do not cancel, we will be getting some information regarding the treatment effects from such a comparison of the blocks. This will be the case with all incomplete block designs.

The process of obtaining some information regarding treatments from such block comparisons is called the "recovery of interblock information" and the information obtained is called inter-block information.

**(iv) Inter-Block Analysis of a BIBD**

Assuming the additive model

$$y_{ij} = \alpha_i + \beta_j + \varepsilon_{ij},$$

the $i-$th block total is given by

$$B_i = k\alpha_i + \sum_j n_{ij}\beta_j + \sum_j \varepsilon_{ij},$$

where $\alpha_i$ is the i-th block effect, $\beta_j$ is the j-th treatment effect, and $\varepsilon_{ij}$ is a random error, which is independently and normally distributed with mean zero and variance $\sigma^2$. Further it is assumed that the block effects $\alpha_i$ are uncorrelated with the treatment effects and the errors. This assumption about the $\alpha_i$'s is validated by the assumption of additivity and the random allocation of groups of treatments to blocks. The sum of the block effects will be

$$B_i - \sum_j n_{ij}\beta_j - \sum_j \varepsilon_{ij}$$

with expected value $B_i - \sum_j n_{ij}\beta_j$, and so the S.S. due to block effects will be

$$L = \sum_i (B_i - \sum_j n_{ij}\beta_j)^2.$$

Minimizing L w.r.t. $\beta_j$, the normal equations are

$$\sum_i n_{ij}(B_i - \sum_j n_{ij}\beta_j)\,\beta_j = 0.$$

$$\begin{aligned}
\text{i.e.,}\quad T_j &= \sum_i n_{ij}(\sum_j n_{ij}\beta_j) \\
&= \beta_1 \sum_i n_{ij}n_{i1} + \cdots + \beta_j \sum_i n_{ij}^2 + \cdots + \beta_v \sum_i n_{ij}n_{iv} \\
&= \lambda\beta_1 + \cdots + \lambda\beta_{j-1} + r\beta_j + \lambda\beta_{j+1} + \cdots + \lambda\beta_v \\
&= \lambda\beta_. + (r-\lambda)\beta_j.
\end{aligned}$$

Summing w.r.t. j, we get

$$\begin{aligned}
\sum_j T_j &= \lambda v\beta_. + (r-\lambda)\beta_. \\
&= [r + \lambda(v-1)]\beta_. \\
&= [r + r(k-1)]\beta_. \\
&= rk\beta_.
\end{aligned}$$

i.e., $kG = rk\beta.$ , where G=grand total.

Therefore

$$\hat{\beta}_. = \frac{G}{r},$$

$$T_j = (r-\lambda)\beta_j + \frac{\lambda}{r}G,$$

and

$$\hat{\beta}_j = \frac{1}{r-\lambda}\left[T_j - \frac{\lambda}{r}G\right].$$

Thus all the v treatment effects are independently estimable and so the minimum S.S. will have

$$\begin{aligned} &\text{d.f.} \\ &= (\text{Total number of observations}) \\ &- (\text{number of indepdnent estimable parameteric functions}) \end{aligned}$$

$$= b - v.$$

The inter-block estimate of the treatment contrast $\beta_i - \beta_j$ is

$$\hat{\beta}_i - \hat{\beta}_j = \frac{1}{r-\lambda}(T_i - T_j).$$

The corresponding intra-block estimate is

$$\hat{\beta}_i - \hat{\beta}_j = Q_i' - Q_j' = \frac{k}{\lambda v}(Q_i - Q_j).$$

The minimum error S.S. due to blocks is

$$\begin{aligned} S'^2 &= \sum_i \left( B_i - \sum_j n_{ij}\hat{\beta}_{ij} \right) B_i \\ &= \sum_i B_i^2 - \sum_i B_i \sum_j n_{ij}\hat{\beta}_{ij}. \end{aligned}$$

This is the S.S. per block. Therefore S.S. per plot is

$$\begin{aligned} S^2 &= \frac{S'^2}{k} \\ &= \frac{1}{k}\left[\sum_i B_i^2 - \sum_i B_i \sum_j n_{ij}\hat{\beta}_j\right] \\ &= \frac{1}{k}\left[\sum_i B_i^2 - \sum_i \hat{\beta}_j \sum_j n_{ij} B_i\right] \end{aligned}$$

$$= \frac{1}{k}\left[\sum_i B_i^2 - \sum_j T_j\hat{\beta}_j\right]$$

$$= \frac{1}{k}\left\{\sum_i B_i^2 - \sum_j T_j\left[\frac{1}{r-\lambda}\left(T_j - \frac{\lambda}{r}G\right)\right]\right\}$$

$$= \frac{1}{k}\left[\sum_i B_i^2 + \frac{\lambda G}{r(r-\lambda)}\sum_j T_j - \frac{1}{r-\lambda}\sum_j T_j^2\right]$$

$$= \frac{1}{k}\left[\sum_i B_i^2 + \frac{k\lambda}{r(r-\lambda)}G^2 - \frac{1}{r-\lambda}\sum_j T_j^2\right]$$

$$= \frac{1}{k}\sum_i B_i^2 - \frac{1}{k(r-\lambda)}\sum_j T_j^2 + \frac{\lambda}{r(r-\lambda)}G^2.$$

This is called the inter-block error S.S.

Now, consider the hypothesis,

$$H_0:\ \beta_1 = \beta_2 = \cdots = \beta_v\ (= \beta,\ \text{say}).$$

The present block error S.S. is

$$L_0 = \sum_i (B_i - k\beta)^2$$

giving the least square normal equations

$$\sum_i (B_i - k\beta) = 0$$

so that $\hat{\beta} = \frac{G}{bk}$, the overall mean. Therefore the conditional minimum is

$$S_0'^2 = \sum_i (B_i - \frac{G}{b})B_i.$$

Therefore the conditional minimum S.S. per plot is

$$S_0^2 = \frac{1}{k}\sum_i (B_i - \frac{G}{b})B_i$$

$$= \frac{1}{k}\sum_i B_i^2 - \frac{G^2}{bk}$$

$$= \frac{1}{k}\sum_i B_i^2 - \frac{G^2}{rv}.$$

(This is the same as the block S.S. ignoring treatments in the intra-block analysis.) Further S.S. due to treatments is

$$S_0^2 - S^2 = \frac{1}{k(r-\lambda)}\sum_j T_j^2 - \left[\frac{\lambda}{r(r-\lambda)} + \frac{1}{rv}\right]G^2$$

$$= \frac{1}{k(r-\lambda)}\left[\sum_j T_j^2 - \frac{k}{v(r-\lambda)}G^2\right]$$

$$= \frac{1}{k(r-\lambda)}\left[\sum_j T_j^2 - \frac{k^2}{v}G^2\right]$$

and it has $(v-1)$ d.f.

**(v) Inter-Block Analysis**

| Source | S.S. | d.f. | M.S.S | F |
|---|---|---|---|---|
| Blocks | $S_0^2 = \frac{1}{k}\sum_i B_i^2 - \frac{G^2}{rv}$ | $b-1$ | | |
| Treatments | $S_0^2 - S^2$ (sub) | $v-1$ | $V_T'$ | $V_T'/V_b$ |
| Error (inter-block) | $S^2$ | $b-v$ | $V_b$ | |

**Inter-Block and Intra-Block Estimates of $\beta_i - \beta_j$:**

The intra-block estimate is

$$T_1 = \frac{k}{\lambda v}(Q_i - Q_j)$$

and the inter-block estimate is

$$T_2 = \frac{1}{r-\lambda}(T_i - T_j).$$

The variances of the two estimates are

$$V_1 = \frac{2k}{\lambda v} V_a \text{ and } V_2 = \frac{2}{r-\lambda} V_b$$

where $V_a = V_E$ is the inter-block error variance and $V_b$ is the inter-block error variance.

Now, as we know, the best unbiased linear combination of $T_1$ and $T_2$ is

$$T = \frac{\frac{1}{V_1}T_1 + \frac{1}{V_2}T_2}{\frac{1}{V_1} + \frac{1}{V_2}}$$

$$= \frac{\frac{Q_i - Q_j}{V_a} + \frac{T_i - T_j}{V_b}}{\frac{\lambda v}{kV_a} + \frac{1}{V_b}}.$$

Since, $Q_i - Q_j = \left(V_i - \frac{1}{k}T_i\right) - \left(V_j - \frac{1}{k}T_j\right)$, we get

$$T = \frac{\left(\frac{V_i - V_j}{V_a}\right) + (T_i - T_j)\left(\frac{kV_a - V_b}{kV_a V_b}\right)}{\frac{\lambda v}{kV_a} + \frac{r-\lambda}{V_b}}.$$

Also, we know the variance of such an estimator is

$$V(T) = \frac{1}{\frac{1}{V_1} + \frac{1}{V_2}} = \frac{2kV_a V_b}{\lambda v V_b + k(r-\lambda)V_a}.$$

**To test $H_0$: $\beta_i = \beta_j$:**

**(i) Based on Intra-Block Analysis**

Under $H_0$,

$$t = \frac{\hat{\beta}_i - \hat{\beta}_j}{\sqrt{\frac{2k}{\lambda v} V_a}}$$

$$= \frac{\frac{k}{\lambda v}(Q_i - Q_j)}{\sqrt{\frac{2k}{\lambda v} V_a}}$$

is a Student's t with $(bk - b - v + 1)$ d.f. The $100(1 - \alpha)\%$ confidence limits for the contrast $\beta_i - \beta_j$ will be

$$(\beta_i - \beta_j) \pm t_{\alpha/2}\sqrt{\frac{2k}{\lambda v} V_a} = \frac{k}{\lambda v}(Q_i - Q_j) \pm t_{\alpha/2}\sqrt{\frac{2k}{\lambda v} V_a}\,.$$

**(ii) Based on Inter-block Analysis**

$$t = \frac{\frac{1}{r - \lambda}(T_i - T_j)}{\sqrt{\frac{2}{r - \lambda} V_b}} \quad \text{with} \quad (b - v) \text{ d.f.}$$

Note:

The inter-block analysis gives meager information regarding the treatments. But the intra-block analysis gives complete information. The intra-block error variance will be less than the inter-block error. Therefore the test based on the intra-block analysis will be the valid one.

## 6.7 Missing Plot

Let the observation corresponding to the j-th treatment in the i-th block be missing. Let $B_i'$ be the sum of the existing observations in the i-th block and $G'$ the grand total of all the known observations. Let X denote the missing value. Then the procedure is to estimate X by minimizing the intra-block error variance. From the ordinary intra-block analysis, we have

$$\text{Total S.S.} = X^2 - \frac{(G' + X)^2}{rv} + \text{terms independent of X}$$

$$\text{Block S.S.} = \frac{(B_i' + X)^2}{k} - \frac{(G' + X)^2}{rv} + \text{terms independent of X.}$$

Now, $Q_j = V_j - \frac{T_j}{k}$, where $V_j$ is the sum of all the observations corresponding to the j-th variety and $T_j$ is the sum of those block totals where the j-th treatment is present. Here the observation corresponding to the j-th variety in the i-th block is missing. So $V_1, V_2, \dots, V_{j-1}, V_{j+1}, \dots, V_v$ will not contain X, and let $V_j'$ be the sum of the existing observations corresponding to the j-th variety. But, because of the definition of $T_j$ we see that X will occur in $T_1$ if the 1-st variety occurs in the i-th block and similarly for $T_2, \dots, T_v$. Of course $T_j$ will contain X and let $T_j = T_j' + X$. Then

$$Q_j = (V_j' + X) - \frac{1}{k}(T_j' + X)$$
$$= \left(V_j' - \frac{1}{k}T_j'\right) + \frac{k-1}{k}X.$$

There are k varieties in the i-th block. Therefore, excluding $T_j$, there will be $(k-1)$ T's in which X is present and correspondingly there will be $(k-1)$Q's in which X is present. Let them be

$$Q_{l_1}, Q_{l_2}, \dots, Q_{l_{k-1}}.$$

Then, $Q_j = \left(V_j' - \frac{1}{k}T_j'\right) + \frac{k-1}{k}X$, and $Q_{l_i}' = \left(V_{l_i}' - \frac{1}{k}T_{l_i}'\right) - \frac{X}{k}$, $i = 1, \dots, k-1$.

Therefore

$$\text{Treatment S.S.} = \frac{k}{\lambda v}\sum_j Q_j^2$$
$$= \frac{k}{\lambda v}\left(Q_j^2 + Q_{l_1}^2 + \cdots + Q_{l_{k-1}}^2\right)$$
$$+ \text{terms independent of X}$$
$$= \frac{k}{\lambda v}\left(Q_j^2 + \sum_{i=1}^{k-1} Q_{l_i}^2\right) + \text{terms independent of X.}$$

Hence, the intra-block error S.S. is

$$E = \text{Total} - \text{Treatment} - \text{Block}$$

$$= X^2 - \frac{(B_i' + X)^2}{k} - \frac{k}{\lambda v}\left(Q_j^2 + \sum_{i=1}^{k-1} Q_{l_i}^2\right)$$

$$+ \text{terms independent of } X.$$

Differentiating w.r.t. X and equating to zero, we get

$$X - \frac{(B_i' + X)}{k} - \frac{k}{\lambda v}\left(\frac{k-1}{k}Q_j + \frac{1}{k}\sum_{i=1}^{k-1} Q_{l_i}\right) = 0. \qquad (3)$$

Now,

$$\frac{k-1}{k}Q_j = \frac{k-1}{k}\left(V_j' - \frac{1}{k}T_j'\right) - \left(\frac{k-1}{k}\right)^2 X,$$

and

$$-\frac{1}{k}Q_{l_i} = -\frac{1}{k}\left(V_{l_i}' - \frac{1}{k}T_{l_i}'\right) + \frac{1}{k^2}X.$$

Therefore

$$\frac{k}{\lambda v}\left(\frac{k-1}{k}Q_j - \frac{1}{k}\sum_{i=1}^{k-1} Q_{l_i}\right)$$

$$= \frac{k-1}{\lambda v}Q_j - \frac{1}{\lambda v}\sum_{i=1}^{k-1} Q_{l_i}$$

$$= \frac{k-1}{\lambda v}\left[\left(V_j' - \frac{1}{k}T_j'\right) + \left(\frac{k-1}{k}\right)X\right] - \frac{1}{\lambda v}\left\{\sum_{i=1}^{k-1}\left[\left(V_{l_i}' - \frac{1}{k}T_{l_i}'\right) - \frac{X}{k}\right]\right\}$$

$$= \frac{X}{k\lambda v}[(k-1)^2 + (k-1)] + \frac{k-1}{\lambda v}\left(V_j' - \frac{T_j}{k}\right)$$

$$- \frac{1}{\lambda v}\sum_{i=1}^{k-1}\left(V_{l_i} - \frac{1}{k}T_{l_i}'\right).$$

Hence from (3),

$$X\left[1 - \frac{1}{k} - \frac{k-1}{\lambda v}\right] = \frac{B_i'}{k} + \frac{k-1}{\lambda v}\left(V_j' - \frac{T_j}{k}\right) - \frac{1}{\lambda v}\sum_{i=1}^{k-1}\left(V_{l_i} - \frac{1}{k}T_{l_i}'\right),$$

giving

$$X = \frac{1}{(k-1)(\lambda v - k)}\left[\lambda v B_i' + k(k-1)\left(V_j' - \frac{T_j}{k}\right) - k\sum_{i=1}^{k-1}\left(V_{l_i} - \frac{1}{k}T_{l_i}'\right)\right]$$

$$= \frac{\lambda v B_i' + kQ_j' - Q}{(k-1)(\lambda v - k)}, \quad \text{on simplification}$$

where

$$Q = \sum_l Q_l', \quad Q_l' = kV_l' - T_l'$$

and where l denotes a variety occurring in the i-th block.

## 6.8 Principle of Connectedness

Often the experimenter is interested in getting unbiased estimates of the treatment effects. For this purpose it is important to know which linear functions of the treatments are estimable. Usually the experimenter will be interested in special types of linear functions, namely contrasts. For example, in ordinary agricultural experiments, the interest will be mainly on treatment effects and not in block effects and consequently the experimenter will be interested in linear functions involving treatment effects only. To answer the question, which linear functions of the treatment effects are estimable, we introduce the idea of connectedness.

The concept of connectedness is based on a chain of relationships. A treatment and a block in a design are said to be associated if the treatment occurs in the block. Two treatments, two blocks, a treatment and a block are said to be connected if it is possible to pass from one to the other by means of a chain consisting alternately of blocks and treatments such that any two members of the chain are associated.

## Example 6.2

Let 1,2,3, ... denote the treatments and let the arrangement be as follows:

$$B_1: \boxed{1|2|3},\quad B_2: \boxed{2|4|5},\quad B_3: \boxed{4|6|8}.$$

Blocks $B_1$ and $B_2$ are connected, for there exists the chain of relationships

$$B_1 \to 2 \to B_2.$$

Treatments 1 and 8 are connected:

$$1 \to B_1 \to 2 \to B_2 \to 4 \to B_3 \to 8.$$

A design (or a portion of a design) is said to be a connected design (or a connected portion of a design) if every block and treatment of it is connected to every other. In general, a design must always break up into a number of connected portions. Then clearly a block or treatment belonging to any one portion is unconnected with a block or treatment belonging to any other potion.

**Example 6.3**

Consider the following layout

$$B_1: \boxed{1|4|6},\ B_2: \boxed{2|3|4},\ B_3: \boxed{3|5|7},\ B_4: \boxed{8|9|10},\ B_5: \boxed{9|11|12}.$$

Treatments 1 and 7 are connected, for

$$1 \to B_1 \to 4 \to B_2 \to 3 \to B_3 \to 7.$$

Similarly treatments 1 and 8 are connected. Blocks $B_1$ and $B_2$, $B_1$ and $B_3$, $B_2$ and $B_3$ are connected. But $B_1, B_2$ and $B_3$ are not connected with $B_4$ and $B_5$, however $B_4$ and $B_5$ are connected. So there are two connected portions in this design. the first connected portion consists of $B_1$, $B_2$ and $B_3$ and the second connected portion consists of $B_4$ and $B_5$.

Consider a design with m connected portions. Let a number of v treatments be tried in b blocks such that

$$v = v_1 + v_2 + \cdots + v_m,$$

$$b = b_1 + b_2 + \cdots + b_m,$$

where $v_i$ is the number of varieties tried in the i-th connected portion and $b_i$ is the number of blocks in the i-th connected portion. Let the total number of plots in the i-th connected portion be $n_i$. Let

$$n = n_1 + n_2 + \cdots + n_m.$$

As the number of plots is n we will get n observational equation. Let the observation vector be

$$y = (y_1, y, \dots, y_m),$$

where

$$y_i = (y_{i1}, y_{i2}, \dots, y_{in_i}).$$

Let the treatment effects be

$$\beta_{11}, \beta_{12}, \dots, \beta_{1v_1}; \beta_{21}, \beta_{22}, \dots, \beta_{2v_2}; \dots, \beta_{m1}, \beta_{m2}, \dots, \beta_{mv_m}.$$

and the block effects be

$$\alpha_{11}, \alpha_{12}, \dots, \alpha_{1b_1}; \alpha_{21}, \alpha_{22}, \dots, \alpha_{2b_2}; \dots, \alpha_{m1}, \alpha_{m2}, \dots, \alpha_{mb_m}.$$

The total number of parameters is $b + v$. Let $T_i$ denote the vector of the treatment parameters in the i-th connected portion and $B_i$ the block parameters in the same. Then

$$T_i = (\beta_{i1}, \beta_{i2}, \dots, \beta_{iv_i}), \quad i = 1, \dots, m,$$
$$B_i = (\alpha_{i1}, \alpha_{i2}, \dots, \alpha_{ib_i}), \quad i = 1, \dots, m.$$

Let

$$\theta = (T_1, \dots, T_m; B_1, \dots, B_m).$$

Then the observational equations become

$$E(y) = \theta A^T,$$

where A is the coefficient matrix. Let $A_i$ be the matrix of the coefficients of the treatments in the i-th connected portion, and $P_i$ the matrix corresponding to the block effects in the same. Then

$$E(y) = (T_1, \dots, T_m; B_1, \dots, B_m)A^T,$$

where

$$A = (A_1, \dots, A_m; P_1, \dots, P_m).$$

Clearly, the number of rows of $A_i^T$ is $v_i$, the number of rows of $P_i^T$ is $b_i$ and the number of columns for both is n, the total number of observations. Here $\theta$ is a $1 \times (b + v)$ matrix and $A^T$ will have $(b + v)$ rows. It will be of the form

$$A^T = \begin{pmatrix} A_1^T \\ \vdots \\ A_m^T \\ P_1^T \\ \vdots \\ P_m^T \end{pmatrix}, \quad A_i^T \to v_i \times n, \quad P_i^T \to b_i \times n.$$

In θ, $T_i$ (Treatments) and $B_i$ (Blocks) will be the parametric portions corresponding to $A_i^T$ and $P_i^T$ respectively. Since the $A_i$ and $P_i$ corresponding to observations in the other connected portions will all be 0, the entries in the first $n_1 + \cdots + n_{i-1}$ and the last $n_{i+1} + \cdots + n_m$ columns in them will all be zero. Let us consider the $n_i$ columns which are non-zero. Further the column corresponding to each observation will be a unit vector because each observation will be based only on a single treatment and single block and so the other coefficients in the linear model will be zero. Thus $A_i^T$ will be of the form

$$A_i^T = \begin{pmatrix} n_1 & n_2 & \ldots n_{i-1} & n_i & n_{i+1} & \ldots & n_m \\ 0 & 0 & \ldots\ 0 & \delta_{11}, \ldots, \delta_{1n_i} & 0 & \ldots & 0 \\ \ldots & \ldots & \ldots\ \ldots & \ldots & \ldots & \ldots & \ldots \\ 0 & 0 & \ldots\ 0 & \delta_{v_i1}, \ldots, \delta_{v_in_i} & 0 & \ldots & 0 \end{pmatrix}$$

where

$\delta_{pq} = 1$, if the
q-th observation in the
i-th connected portion
corresponds to the p-th treatment in that portion
$= 0$, otherwise,

and $P_i^T$ will be of the form

$$P_i^T = \begin{pmatrix} n_1 & n_2 & \ldots n_{i-1} & n_i & n_{i+1} & \ldots & n_m \\ 0 & 0 & \ldots\ 0 & \delta_{11}, \ldots, \delta_{1n_i} & 0 & \ldots & 0 \\ \ldots & \ldots & \ldots\ \ldots & \ldots & \ldots & \ldots & \ldots \\ 0 & 0 & \ldots\ 0 & \delta_{b_j1}, \ldots, \delta_{b_jn_i} & 0 & \ldots & 0 \end{pmatrix}$$

where

$\delta_{pq} = 1$, if the q-th observation in the
i-th connected portion corresponds to the
p-th block in that portion,
$= 0$, otherwise.

Therefore, if we sum all the $v_i$ rows of $A_i^T$, we will get a vector of the form

$$\begin{pmatrix} n_1 & n_2 & \cdots & n_i & \cdots & n_{m-1} & n_m \\ 0,\ldots,0 & 0,\ldots,0 & \ldots & 1,1,\ldots,1 & \ldots & 0,\ldots,0 & 0,\ldots,0 \end{pmatrix}.$$

Similarly, the sum of the $b_i$ rows of $P_i^T$ will be

$$\begin{pmatrix} n_1 & \cdots & n_{i-1} & n_i & n_{i+1} & \cdots & n_m \\ 0,\ldots,0 & \ldots & 0,\ldots,0 & 1,1,\ldots,1 & 0,\ldots,0 & \ldots & 0,\ldots,0 \end{pmatrix}.$$

Therefore

$$\text{sum of the } v_i \text{ rows of } A_i^T = \text{ sum of the } b_i \text{ rows of } P_i^T.$$

This is true for all $i = 1,2,\ldots,m$. Therefore the $(b+v)$ rows of $A^T$ satisfy m linear restrictions and so rank of A = $(b+v-m)$.

Now, consider a parametric function involving treatments alone, say

$$C_{11}\beta_{11} + \cdots + C_{1v_1}\beta_{1v_1} + \cdots + C_{m1}\beta_{m1} + \cdots + C_{mv_m}\beta_{mv_m}.$$

i.e., $\theta C^T$, where $C = (C_1,\ldots,C_m,0,\ldots,0)$ where $C_i = (C_{i1},\ldots,C_{iv_i})$. The coefficients of $B_1,\ldots,B_m$ being 0. Then we know that the necessary and sufficient condition for the estimability of $\theta C^T$ is that A and A augmented with C are of the same rank. But we have seen that Rank $A = b+v-m$ and the rows of $A^T$ satisfy m linear restrictions. Hence, if C augmented with A does not alter the rank of A it is necessary and sufficient that the rows of $C^T$, i.e., the columns of C, also satisfy the same linear restrictions, i.e., the sum of the coefficients in $C_i$ must be equal to zero (which is the sum of the coefficients of the corresponding block effects) for all i. Thus we have the following:

**Theorem 6.5**

The Necessary and Sufficient condition for the estimability of a linear parametric function involving treatments alone is that it should be a contrast with respect to each and every connected portion.

Analysis of Variance in the case of a connected design with m connected portions will be

| Source | d.f. |
|---|---|
| Blocks (ignoring treatments) | $b - 1$ |
| Treatments (eliminating blocks) | $v - m$ |
| Error | $n - (b + v - m)$ |
| Total | $n - 1$ |

# Chapter 7

# Factorial Designs

## 7.1 Introduction

Suppose we are studying the effect of a single treatment only. Then we conduct the experiment with that treatment at various levels and study the variations in the treatment effects as the level increases. If the treatment is manure, say superphosphate, then we apply 50 lbs. in one plot, 100 lbs. in another plot, 150 lbs. in a 3-rd plot and so on. These are called the various levels. Factorial design is in fact a treatment design where we study a treatment in detail, by dividing the total variation into components corresponding to various levels.

Consider a manure $M_1$ at various levels. These different levels can also be taken as various treatments. From such treatment comparisons we can say whether the yield increases or decreases as the amount of manure applied increases. The experiment may be a randomized block or Latin square. Taking the levels as treatments, the data can then be analyzed in the usual way.

Consider two treatments applied at two levels. Let the first treatment be superphosphate, the two levels being 50 lbs./plot and 100 lbs./plot. Let the second treatment be depth of soil, the levels being 5" depth and 10" depth. Considered as a simple design, here there are 4 treatments. But if we consider it as a factorial design, there are only two treatments at two levels each.

Consider a simple factorial experiment with two treatments at two levels each. Let the first treatment levels be $n_0, n_1$ and the second treatment levels be $k_0, k_1$. We will be getting observations corresponding to $n_0k_0, n_0k_1, n_1k_0, n_1k_1$. This is called a $2 \times 2$ or $2^2$ factorial experiment. The index stands for the number of treatments and the base for the number of levels. Consider the following example:

| | $n_0$ | $n_1$ |
|---|---|---|
| $k_0$ | 40 | 45 |
| $k_1$ | 42 | 48 |

Consider the difference

$$n_0k_1 - n_0k_0 = 42 - 40 = 2.$$

This difference is due to the level difference in the second treatment. Again,

$$n_1k_1 - n_1k_0 = 48 - 45 = 3$$

is also the difference due to the level difference in the second treatment. Similarly, the differences due to the level difference in the first treatment are

$$n_1k_0 - n_0k_0 = 45 - 40 = 5$$

$$n_1k_1 - n_0k_1 = 48 - 42 = 6.$$

Thus the simple effects in both cases are different. The difference may be due to experimental error or due to some relationship between the level differences of the two treatments. The second cause is called interaction.

The interaction between two level differences in the case of a factorial design means that the effect of one treatment differs when the level of the other treatment is altered. Let us estimate the simple effects.

$$\frac{1}{2}[(n_1k_1 - n_0k_1) + (n_1k_0 - n_0k_0)]$$

is called the average effect of the level difference in the first treatment. This is also called the main effect of the first treatment. Let the first treatment be denoted by N and the second by K. Then

$$\frac{1}{2}[(n_1k_1 - n_0k_1) + (n_1k_0 - n_0k_0)]$$

is the main effect of N.
Similarly, main effect of K is

$$\frac{1}{2}[(n_1k_1 - n_1k_0) + (n_0k_1 - n_0k_0)].$$

Denoting by N and K the effects of N and K, we may take

$$\hat{N} = \frac{1}{2}[(n_1k_1 - n_0k_1) + (n_1k_0 - n_0k_0)]$$

$$\hat{K} = \frac{1}{2}[(n_1k_1 - n_1k_0) + (n_0k_1 - n_0k_0)].$$

In our example,

$$\hat{N} = \frac{11}{2} \text{ and } \hat{K} = \frac{5}{2}.$$

Let us estimate the interaction. The difference between

$$(n_0k_1 - n_0k_0) \text{ and } (n_1k_1 - n_1k_0)$$

will give the interaction component of K to N, denoted by KN. Usually $n_1$ is a higher level than $n_0$ and then we take the difference as

$$(n_1k_1 - n_1k_0) - (n_1k_0 - n_0k_0).$$

If it is positive, then the interaction is increasing as the level increases. Then

$$\widehat{KN} = (n_1k_1 - n_1k_0) - (n_0k_1 - n_0k_0).$$

Similarly,

$$\widehat{NK} = (n_1k_1 - n_0k_1) - (n_1k_0 - n_0k_0).$$

But NK=KN. The overall mean is

$$M = \frac{1}{4}(n_0k_0 + n_1k_0 + n_0k_1 + n_1k_1).$$

Let us form the following table

| | $n_0k_0$ | $n_0k_1$ | $n_1k_0$ | $n_1k_1$ | Dividing factors |
|---|---|---|---|---|---|
| N | $-1$ | $-1$ | 1 | 1 | 2 |
| K | $-1$ | 1 | $-1$ | 1 | 2 |
| NK | 1 | $-1$ | $-1$ | 1 | 2 |
| M | 1 | 1 | 1 | 1 | 4 |

All the four vectors are mutually orthogonal. Hence all the effects are mutually orthogonal.

$$\hat{N} = \frac{1}{2}(n_1k_0 - n_0k_0 + n_1k_1 - n_0k_1)$$

$$= \frac{1}{2}(n_1 - n_0)(k_1 + k_0)$$

$$\hat{K} = \frac{1}{2}(n_1k_1 - n_1k_0 + n_0k_1 - n_0k_0)$$

$$= \frac{1}{2}(n_1 + n_0)(k_1 - k_0)$$

$$\widehat{NK} = \frac{1}{2}[(n_1k_1 - n_1k_0) - (n_0k_1 - n_0k_0)]$$

$$= \frac{1}{2}(n_1 - n_0)(k_1 - k_0).$$

There are 4 treatments and 3 components of variation, viz. N, K and NK. Hence the total variation can be divided into variation due to N, variation due to K and variation due to NK. Altogether there are 3 d. f. which can be divided into one each.

Let us consider a randomized block experiment with 5 replications of these 4 treatments. There will be 5 blocks and 4 plots in each block. The 4 treatments are applied to the plots of a block at random.

$$\text{Total d.f.} = 19.$$

$$\text{d.f. due to blocks} = 4$$

$$\text{d.f. due to error} = 12.$$

Let

$$n_0k_0 = t_1$$

$$n_1k_0 = t_2$$

$$n_0k_1 = t_3$$

$$n_1k_1 = t_4.$$

Then

$$\hat{N} = \frac{1}{2}[t_2 + t_4 - t_1 - t_3]$$

and

$$\hat{K} = \frac{1}{2}[-t_1 - t_2 + t_3 + t_4].$$

If there are r replications, the dividing factor is $2r$. Then $n_0k_0$ will correspond to a sum of r observations. We get

$$\hat{N} = \frac{1}{2r}[-t_1 + t_2 - t_3 + t_4],$$

$$\hat{K} = \frac{1}{2r}[-t_1 - t_2 + t_3 + t_4],$$

$$\widehat{NK} = \frac{1}{2r}[t_1 - t_2 - t_3 + t_4],$$

$$\hat{M} = \frac{1}{4r}[t_1 + t_2 + t_3 + t_4].$$

Then

$$\begin{aligned} t_1 &= \frac{4rM + 2rNK - 2rK - 2rN}{4} \\ &= r\left[M + \frac{NK}{2} - \frac{K}{2} - \frac{N}{2}\right], \\ t_2 &= r\left[M - \frac{NK}{2} - \frac{K}{2} + \frac{N}{2}\right], \\ t_3 &= r\left[M - \frac{NK}{2} + \frac{K}{2} - \frac{N}{2}\right], \\ t_4 &= r\left[M + \frac{NK}{2} + \frac{K}{2} + \frac{N}{2}\right]. \end{aligned}$$

Let T denote the vector $(t_1, t_2, t_3, t_4)$. Then

$$\begin{aligned} T &= r(M \quad N \quad K \quad NK)\begin{pmatrix} 1 & 1 & 1 & 1 \\ -\frac{1}{2} & \frac{1}{2} & -\frac{1}{2} & \frac{1}{2} \\ -\frac{1}{2} & -\frac{1}{2} & \frac{1}{2} & \frac{1}{2} \\ \frac{1}{2} & -\frac{1}{2} & -\frac{1}{2} & \frac{1}{2} \end{pmatrix} \\ &= r\theta A \end{aligned}$$

where $\theta = (M \quad N \quad K \quad NK)$ and

$$A = \begin{pmatrix} 1 & 1 & 1 & 1 \\ -\frac{1}{2} & \frac{1}{2} & -\frac{1}{2} & \frac{1}{2} \\ -\frac{1}{2} & -\frac{1}{2} & \frac{1}{2} & \frac{1}{2} \\ \frac{1}{2} & -\frac{1}{2} & -\frac{1}{2} & \frac{1}{2} \end{pmatrix}.$$

S.S. of the observations is

$$\sum t_i^2 = TT^T = r^2\theta AA^T\theta^T$$

$$= r^2\theta \begin{pmatrix} 1 & 0 & 0 & 0 \\ 0 & 1 & 0 & 0 \\ 0 & 0 & 1 & 0 \\ 0 & 0 & 0 & 1 \end{pmatrix} \theta^T.$$

$$= r^2[4M^2 + N^2 + K^2 + (NK)^2].$$

Therefore

$$\frac{\sum t_i^2}{r} - 4rM^2 = r[N^2 + K^2 + (NK)^2].$$

In terms of the grand total G, $M = \frac{G}{4r}$. Therefore

$$4rM^2 = \frac{G^2}{4r}.$$

Thus $\frac{\sum t_i^2}{r} - \frac{G^2}{4r}$ is the variation corresponding to the treatments.

$$\text{S.S. due to N} = rN^2 = \frac{1}{4r}(-t_1 + t_2 - t_3 + t_4)^2$$

$$\text{S.S. due to K} = rK^2 = \frac{1}{4r}(-t_1 - t_2 + t_3 + t_4)^2$$

$$\text{S.S. due to NK} = r(NK)^2 = \frac{1}{4r}(t_1 - t_2 - t_3 + t_4)^2$$

Therefore

$$\text{Total S.S. due to treatments}$$
$$= \text{S.S. due to N} + \text{S.S. due to K} + \text{S.S. due to NK}.$$

A test for the interaction effect will be

$$F(1, n_e) = \frac{\text{S.S. due to NK}/1}{\text{Error S.S.}/n_e} = \frac{\text{S.S. due to NK}}{V_E}$$

where $V_E$ = error mean square in the ordinary analysis, and

$$n_e = \text{d.f. due to error.}$$

In the above particular case, $n_e = 12$. Similarly, a test for the variation due to N will be

$$F(1, n_e) = \frac{\text{S.S. due to N}}{V_E}$$

and a test for the variation due to K will be

$$F(1, n_e) = \frac{\text{S.S. due to K}}{V_E}.$$

**Yates' method for getting the various S.S.**

Arrange the treatments in the form $n_0k_0, n_1k_0, n_0k_1, n_1k_1$ such that the level will be increased step by step. This we take as the 1-st column. In the second column we write $t_1, t_2, t_3, t_4$. Fill the other columns as shown. The dividing factor will be given in (6), r is the number of replications, 4 will be the number of treatments.

| (1) | (2) | (3) | (4) | (5) S.S. | (6) |
|---|---|---|---|---|---|
| $n_0k_0$ | $t_1$ | $t_1 + t_2$ | $t_1 + t_2 + t_3 + t_4$ | $M^2$ | $r \times 4$ |
| $n_1k_0$ | $t_2$ | $t_3 + t_4$ | $-t_1 + t_2 - t_3 + t_4$ | $N^2$ | $r \times 4$ |
| $n_0k_1$ | $t_3$ | $t_2 - t_1$ | $-t_1 - t_2 + t_3 + t_4$ | $K^2$ | $r \times 4$ |
| $n_1k_1$ | $t_4$ | $t_4 - t_3$ | $t_1 - t_2 - t_3 + t_4$ | $(NK)^2$ | $r \times 4$ |

**(i) Sum Check**

The 1-st element in the 4-th column will be the sum of all the observations, i.e., the grand total. The sum of all the observations in the 4-th column is

$$4t_4 = t_4 \times (\text{total number of treatment combinations})$$
$$= t_4 \times (\text{total number of rows}).$$

**(ii) Square Sum Check**

The S.S. of all the observations in the 4-th column will be $(t_1^2 + t_2^2 + t_3^2 + t_4^2) =$ sum of squares of the elements in the second column.

Change in Nation

1 will denote all the factors appearing in the lowest level.

n will denote the factor N appearing in the highest level and all other factors in the lowest level. Similarly for other factors also.

nk will denote the factors N, K appearing at the highest levels and all other factors in the lowest levels.

Let there be 3 factors N, K, P appearing at 2 levels each, i.e., a $2^3$ factorial experiment. Then in the Yates' method where the levels of factors will be increased step by step, the arrangement will be (3 factors at 2 levels each)

1
n
k
nk
p
np
kp
nkp

In general, in a $2^n$ factorial arrangement there will be $2^n$ treatment combinations resulting in $2^n$ rows in Yates process. In a $p^n$ factorial arrangement, there will be $p^n$ treatment combinations.

In one new notation, $n-1$ will be a simple effect of N. ($n-1 = n_1k_0 - n_0k_0$). Similarly, $nk - k$ will be a simple effect of N.

Let there be 3 factors N, K, P at 2 levels. Then

$$n - 1 = n_1k_0p_0 - n_0k_0p_0 = \text{simple effect of N.}$$

$$nk - k = n_1k_1p_0 - n_0k_1p_0 = \text{simple effect of N.}$$

$$np - p = n_1k_0p_1 - n_0k_0p_1 = \text{simple effect of N.}$$

$$npk - k = n_1k_1p_1 - n_0k_1p_1 = \text{simple effect of N.}$$

We have exhausted all possible simple effects for N. Then the main effect of N will be estimated by a simple average of these 4 simple effects.

$$\begin{aligned}N &= \frac{1}{4}[(n-1) + (nk-k) + (np-p) + (npk-pk)]\\ &= \frac{1}{4}[(n-1) + (n-1)k + (n-1)p + (n-1)pk]\\ &= \frac{1}{4}[(n-1)(1+k+p+pk)]\\ &= \frac{1}{4}(n-1)(p+1)(k+1).\end{aligned}$$

Similarly, the main effects of K and P will be

$$K = \frac{1}{4}(n+1)(p+1)(k-1)$$

and

$$P = \frac{1}{4}(n+1)(p-1)(k+1).$$

Now, $n-1$ and $np-p$ are simple effects of N. If there is a difference between these, then that difference is due to the interaction NP between N and P when K appears at the lowest level. Thus

$$(np - p) - (n - 1) = \text{simple effect of NP,}$$

i.e., interaction between N and P when K appears at the lowest level. Similarly,

$$(npk - pk) - (nk - k) = \text{simple effect of NP},$$

i.e., interaction NP when K appears at the highest level.

Similarly, we will get two more expressions, but they will be the same.

$$\begin{aligned} NP &= \frac{1}{4}[(np - p) - (n - 1) + (npk - pk) - (nk - k)] \\ &= \frac{1}{4}(n - 1)(p - 1)(k + 1), \end{aligned}$$

and,

$$NK = \frac{1}{4}(n - 1)(p + 1)(k - 1).$$

$$PK = \frac{1}{4}(n + 1)(p - 1)(k - 1).$$

NP, NK and PK are called the first order interactions.

Now, if there is a difference between $(npk - pk) - (nk - k)$ and $(np - p) - (n - 1)$, then that difference is due to the interaction between NP and K, denoted by NPK.

$$\begin{aligned} NPK &= PNK = NKP \\ &= \{[(npk - pk) - (nk - k)] - [(np - p) - (n - 1)]\} \\ &= \frac{1}{4}(n - 1)(p - 1)(k - 1). \end{aligned}$$

NPK is called a second order interaction. The overall mean will be

$$M = \frac{1}{8}(1 + n + p + k + np + pk + nk + npk).$$

The signs of the various expressions can be arranged as below:

| | $t_1$<br>1 | $t_2$<br>n | $t_3$<br>p | $t_4$<br>np | $t_5$<br>k | $t_6$<br>nk | $t_7$<br>pk | $t_8$<br>npk |
|---|---|---|---|---|---|---|---|---|
| N | −1 | 1 | −1 | 1 | −1 | 1 | −1 | 1 |
| P | −1 | −1 | 1 | 1 | −1 | −1 | 1 | 1 |
| K | −1 | −1 | −1 | −1 | 1 | 1 | 1 | 1 |
| NP | 1 | −1 | −1 | 1 | 1 | −1 | −1 | 1 |
| NK | 1 | −1 | 1 | −1 | −1 | 1 | −1 | 1 |
| PK | 1 | 1 | −1 | −1 | −1 | −1 | 1 | 1 |
| NPK | −1 | 1 | 1 | −1 | 1 | 1 | −1 | −1 |
| M | 1 | 1 | 1 | 1 | 1 | 1 | 1 | 1 |

It can be seen that all the coefficient vectors are mutually orthogonal, i.e., all the effects are contrasts. Therefore, any effect can be written in the form

$$\sum a_i y_i, \text{ where } a_i = \pm 1 \text{ and } \sum a_i = 0.$$

In a $2^n$ factorial design it is not possible to write down such a table for signs. Suppose we are given an effect which contains an even number of factors, such as NP, NK etc. We want to find the main effect corresponding to this effect. If there is an even number of elements corresponding to a treatment combination and the particular effect under consideration, then the sign of the treatment combination is +1.

If there is an odd number of elements common to a treatment combination and the particular effect under consideration then the sign of the treatment combination is +1.

Next, let us consider an effect with an odd number of factors, say N, P, K, NPK etc. If there is an odd number of elements common to a treatment combination and the particular effect, then the sign will be positive. If there is an even number of elements common to a treatment combination and the particular effect then the sign of the treatment combination is negative.

These are the general rules to find the signs of treatment combinations and they can be verified in the case of the above table for a $2^3$ factorial design.

In a $2^n$ factorial experiment, the $2^n$ elements will form a group w.r.t. multiplication, if we assume that the inverse of every element is itself. [In a $2^3$ experiment, if we consider the set of 8 elements and if we assume that the square of each element is unity, i.e., $n^2 = 1$, $k^2 = 1$ etc., then the set of elements is closed under multiplication, the set is associative, the inverse of any element is itself and the set is commutative also. Thus the elements form an abelian group.]

In a $2^3$ factorial experiment, let the treatment combinations be denoted by $t_1, t_2, \ldots, t_8$. Let

$$T = (t_1, t_2, \ldots, t_8)$$

$$E = (M, N, P, K, NP, NK, PK, NPK)$$

Then $E = TA$, where

$$A = \begin{pmatrix} 1 & 1 & 1 & 1 & 1 & 1 & 1 & 1 \\ -1 & 1 & -1 & 1 & -1 & 1 & -1 & 1 \\ -1 & -1 & 1 & 1 & -1 & -1 & 1 & 1 \\ -1 & -1 & -1 & -1 & 1 & 1 & 1 & 1 \\ 1 & -1 & -1 & 1 & 1 & -1 & -1 & 1 \\ 1 & -1 & 1 & -1 & -1 & 1 & -1 & 1 \\ 1 & 1 & -1 & -1 & -1 & -1 & 1 & 1 \\ -1 & 1 & 1 & -1 & 1 & -1 & -1 & 1 \end{pmatrix}.$$

Then $EE^T = TAA^TT^T = 8TT^T = 8\sum_i t_i^2$.

Therefore

$$\sum_i t_i^2 = \frac{1}{8}EE^T$$

$$= \frac{1}{8}[M^2 + N^2 + P^2 + K^2 + (NP)^2 + (NK)^2 + (PK)^2 + (NPK)^2].$$

If there are r replications,

$$\frac{\sum_i t_i^2}{r} = \frac{1}{8r}[M^2 + N^2 + P^2 + K^2 + (NP)^2 + (NK)^2 + (PK)^2 + (NPK)^2].$$

Therefore

$$\sum_i \frac{t_i^2}{r} - \frac{M^2}{8r} = \frac{N^2 + P^2 + K^2 + \cdots (NPK)^2}{8r}.$$

This is the total variation due to treatments and it can be divided into variation due to the different components such as

Variation due to N,

Variation due to P,

Variation due to K,

Variation due to NP,

Variation due to NK,

Variation due to PK,

Variation due to NPK.

The treatment S.S. has 7 d. f. and each of the 7 components on the right has 1 d. f. so that we have a corresponding splitting of the d. f. also.

In the case of a $2^3$ factorial experiment, the Yate's table will be as follow:

| (1) | (2) | (3) | (4) | (5) | (6) |
|---|---|---|---|---|---|
| 1 | $t_1$ | $S_1$ | $S_1'$ | $S_1''$ | $M^2$ |
| n | $t_2$ | $S_2$ | $S_2'$ | $S_2''$ | $N^2$ |
| p | $t_3$ | $S_3$ | $S_3'$ | $S_3''$ | $P^2$ |
| np | $t_4$ | $S_4$ | $S_4'$ | $S_4''$ | $(NP)^2$ |
| k | $t_5$ | $D_1$ | $D_1'$ | $D_1''$ | $K^2$ |
| nk | $t_6$ | $D_2$ | $D_2'$ | $D_2''$ | $(NK)^2$ |
| pk | $t_7$ | $D_3$ | $D_3'$ | $D_3''$ | $(PK)^2$ |
| npk | $t_8$ | $D_4$ | $D_4'$ | $D_4''$ | $(NPK)^2$ |

Consider a separation of the form shown in columns (1) and (2), i.e., divide by the number of levels

$$S_1 = t_1 + t_2,\ S_2 = t_3 + t_4,\ S_3 = t_5 + t_6,\ S_4 = t_7 + t_8;$$
$$D_1 = t_2 - t_1,\ D_2 = t_4 - t_3,\ D_3 = t_6 - t_5,\ D_4 = t_8 - t_7;$$
$$S_1' = S_1 + S_2,\ S_2' = S_3 + S_4,\ S_3' = D_1 + D_2,\ S_4' = D_3 + D_4;$$
$$D_1' = S_2 - S_1,\ D_2' = S_4 - S_3,\ D_3' = D_2 - D_1,\ D_4' = D_4 - D_3;$$
$$S_1'' = S_1' + S_2',\ S_2'' = S_3' + S_4',\ S_3'' = D_1' + D_2',\ S_4'' = D_3' + D_4';$$
$$D_1'' = S_2' - S_1',\ D_2'' = S_4' - S_3',\ D_3'' = D_2' - D_1',\ D_4'' = D_4' - D_3'.$$

We may stop here, since we get the sum of all the observations in $S_1''$. In general, we repeat the operations as many times as there are factors and subdivision will be according to the number of levels. If there are 3 levels we separate 3 by 3.

Dividing factor will be $8r$. Thus

$$\text{S.S. due to N} = \frac{S_2''^2}{8r},$$

$$\text{S.S. due to P} = \frac{S_3''^2}{8r}, \quad \text{etc.}$$

Sum checks.

$$\text{Sum of (5)} = 8t_8$$

$$\text{S.S. of (5)} = 8 \times \text{S.S. of (2)}, \quad \text{etc.}$$

The F-ratios will test the various effects.

## 7.2 Generalization

### $2^n$ Factorial Design

There will be n factors appearing at 2 levels each and there will be $2^n$ treatment combinations. For convenience denote by $A_1, A_2, \dots, A_n$ the factors and by $a_i$ the treatment combination in which $A_i$ appears at the highest level and all other factors at the lowest level. As before, 1 will stand for all the treatments appearing at the lowest level. Similarly, $a_i a_j$ will denote the treatment combination in which $A_i$ and $A_j$ appear at the highest level and all others at the lowest level.

Consider 2 treatment combinations, say $x = a_1a_2a_3$, $y = a_2a_3a_5$. Let us define the product $xy$ as $xy = a_1a_5$, i.e., in the product we delete all common elements, which is equivalent to assuming that the square of any element is unity.

Consider the $2^n$ treatment combinations as elements of a set S. Then, with multiplication as defined above, S is a group of order $2^n$.

Let g denote the set of all treatment combinations which contain a particular element $a_i$. Let $g'$ be the complement of g. Consider $a_ig'$. All elements are distinct from those of $g'$ and all the elements contain $a_i$. All the elements are distinct also and the number of elements will be the same as in $g'$. Hence all the elements of $a_ig'$ belong to g. Therefore the number of elements where $a_i$ is present is greater than or equal to the number of elements where $a_i$ is not present. Consider also the product $a_ig$. No element will contain $a_i'$. So the number of treatment combinations where $a_i$ is not present will be greater than or equal to the number of treatment combinations where $a_i$ is present.

From above, it follows that the number of treatment combinations where $a_i$ is present is equal to the number of treatment combinations where $a_i$ is not present. Therefore, out of the $2^n$ elements, $2^{n-1}$ elements belong to g and $2^{n-1}$ elements belong to $g'$. Evidently, g cannot form a group since it is not closed. But $S = g \cup g'$ and the two sets g and $g'$ are isomorphic. The isomorphism is defined by

$$x \to a_ix, \quad x \in g.$$

x is a treatment combination where $a_i$ is present and $a_ix$ is a treatment combination where $a_i$ is not present. So all the simple effects of $A_i$ can be obtained by subtracting the elements of $g'$ from the corresponding elements of g. So the main effect of $A_i$ will be given by the sum of all the elements of $g - g'$ divided by $2^{n-1}$.

Let G denote the set of all treatment combinations where neither $a_i$ nor $a_j$ is present. Then, clearly

$$S = G \cup a_iG \cup a_jG \cup a_ia_jG,$$

and the number of elements in each is the same. Thus S can be partitioned into 4 mutually exclusive sets of $\frac{2^n}{4} = 2^{n-2}$ elements each. Now, $a_iG - G$ gives the simple effects of $A_i$ in the absence of $A_j$ and

$a_ia_jG - a_jG$ gives the simple effects of $A_i$ in the presence of $A_j$. In general, they will be different and the different elements in

$$(a_ia_jG - a_jG) - (a_iG - G) = (a_ia_jG + G) - (a_iG + a_jG)$$

will measure the interaction between $A_i$ and $A_j$ denoted by $A_iA_j$. So, if r is the number of the replications, the sum of all the elements in $(a_ia_jG + G) - (a_iG + a_jG)$ divided by $2^{n-1}r$ will give the interaction component of $A_iA_j$. Similarly for other interactions.

Let X be a particular effect and x a particular treatment combination. If there is an even number of elements in common between X and x then the sign of x will be positive. Suppose x does not contain $a_i$. Then $a_ix$ will contain $a_i$ and if X also contains $A_i$, then the sign will be negative.

Suppose x does not contain $a_j$ also, but X contains $A_j$ also. Then the signs are as follows:

| | |
|---|---|
| x | + |
| $a_ix$ | − |
| $a_jx$ | − |
| $a_ia_jx$ | + |

The general situation will be as follows:

Let g be a particular treatment combination and X an effect which contains $A_i$ and $A_j$. If there is an even number of elements common between g and X then

| | |
|---|---|
| the sign of g will be | + ve |
| the sign of $a_ig$ will be | − ve |
| the sign of $a_jg$ will be | − ve |
| the sign of $a_ia_jg$ will be | + ve. |

All these treatment combinations are elements in S. The effect of $A_iA_j$ can be obtained by summing the 4 elements.

Consider two effects X and Y. Let $A_i$ be present in X and $A_j$ be absent in X. Let $A_j$ be present in Y and $A_i$ be absent in Y.

Let g be a set of treatment combinations which has an even number of elements common to X as well as to Y. Then g is even with X and Y and $a_i$ and $a_j$ will be present, in g.

$a_i g$ will be odd with X and even with Y. Similarly, $a_j g$ will be odd with Y and even with X. Consider $a_i a_j g$. It will have an even number of elements common to X and odd with Y. Also we know that the $2^n$ elements can be partitioned into the 4 sets $g, a_i g, a_j g$ and $a_i a_j g$. So we can form a table of the form

| X \ Y | Even | Odd |
|---|---|---|
| Even | $g$ | $a_j g$ |
| Odd | $a_i g$ | $a_i a_j g$ |

**The Interaction XY**

If X and Y are two effects, then their product XY is defined as the effect obtained after cancelling the common factors.

**Example 7.1**

Let $X = A_1A_2A_3$, and $Y = A_2A_3A_4A_5$, then $XY = A_1A_4A_5$. It is equivalent to saying that the square of each factor is unity.

Let g be a set of elements which have an even number of elements common to X and Y. Let $2m$ and $2m'$ be respectively the number of elements common between X and g and between Y and g respectively. Let the number of factors in X be $r_1 + r$ and in Y be $r_2 + r$ such that there are r factors common to X and Y. Let s denote the number of elements common to X and Y and at the same time common to g also.

**Example 7.2**

Let $g = \{a_1a_2a_6a_7\}$, $X = A_1A_2A_3$, and $Y = A_2A_3A_4A_5$.
Here $2m = 2$,
$2m' = 2$, $r_1 = 0$, $r = 3$, $r_2 = 1$, and $s = 2$.

In the product XY, the r common elements will be cancelled and there will be $2m - s$ elements common to the remaining elements in X and $g$. Similarly, there will be $2m' - s$ elements common to the remaining elements in Y and $g$. Therefore the number of elements common to XY and $g$ will be

$$(2m - s) + (2m' - s) = 2(m + m') - 2s,$$

which again is even. The product XY formed in this way is defined as the interaction between X and Y.

**Generalization**

Let the number of elements common to $g$ and X be $m_1$ and common to $g$ and Y be $m_2$. Let the number of factors in X be $r_1 + r$ and in Y be $r_2 + r$ where there are r factors common to X and Y. Let s be the number of elements common to X, Y and $g$. Proceeding as before the number of elements common to X Y and $g$ will be $m_1 + m_2 - 2s$. If $m_1$ and $m_2$ are odd, then $m_1 + m_2 - 2s$ will be even.

Now, if $m_1$ is odd and $m_2$ is even or $m_1$ is even and $m_2$ is odd, then $m_1 + m_2 + 2s$ is odd.

Consider, again, a $2^n$ factorial experiment. Any effect can be considered as an interaction. All interactions involving one factor alone are called 0-th order interactions. Interactions involving r factors are called $(r - 1)$-th order interactions. There are $\binom{n}{r}$ $(r - 1)$-th order interactions. The total d. f. is $2^n-1$ which can be divided into single d. f. corresponding to the various interactions. Yates' table for this experiment will be as follows.

| (1) | (2) | (3) | (4) | (5) | (6) |
|---|---|---|---|---|---|
| 1 | $t_1$ | $S_1$ | $S'_1$ | ... | ... |
| $a_1$ | $t_2$ | $S_2$ | $S'_2$ | ... | ... |
| $a_2$ | $t_3$ | $S_3$ | $S'_3$ | ... | ... |
| $a_1a_2$ | $t_4$ | $S_4$ | | ... | ... |
| $a_3$ | $t_5$ | $S_5$ | ... | ... | ... |
| $a_1a_3$ | $t_6$ | $S_6$ | ... | ... | ... |
| $a_2a_3$ | $t_7$ | ... | ... | ... | ... |
| $a_1a_2a_3$ | $t_8$ | ... | ... | ... | ... |
| $a_4$ | $t_9$ | ... | ... | ... | ... |
| $a_1a_4$ | $t_{10}$ | ... | ... | ... | ... |
| $a_2a_4$ | $t_{11}$ | ... | ... | ... | ... |
| $a_1a_2a_4$ | $t_{12}$ | ... | ... | ... | ... |
| ... | ... | ... | ... | ... | ... |

Repeat the operations as many times as there are factors. Note that the 3-rd column is the result of the 1-st operation, 4-th is the result of the 2-nd operation, etc.

The dividing factor will be $2^n \times r$ where r is the number of replications. Let the last column be denoted by $T_0, T_1, T_2, \ldots$, then

$$\text{S.S. due to } A_1 = \frac{T_1^2}{2^n r} \text{ with one d.f.}$$

$$\text{S.S. due to } A_2 = \frac{T_2^2}{2^n r} \text{ with one d.f. etc}$$

For testing all these effects the F-ratio will be $F(1, n_e)$ where $n_e$ is the d. f. of the error.

## 7.3 Confounding in Factorial Designs

Consider a $2^{10}$ factorial design. It is impossible to find enough homogeneous blocks to try all these $2^{10} = 1024$ treatment combinations. There are several suggestions.

(i) Avoid replications. But this is not acceptable as the information available from such an experiment is meager.

(ii) Fractional replication — avoiding those treatment

combinations in which the experimenter is not interested as also higher order interactions.

(iii) Confounded Design. The necessity arises when

(a) the number of treatment combinations is large.

(b) homogeneous blocks are not easily available

(c) higher order interactions cannot be explained or the experimenter is not interested in certain treatment combinations.

Let $T_1, T_2, T_3, T_4$ be 4 treatments with effects $t_1, t_2, t_3, t_4$, tried in two blocks $B_1, B_2$ and let the yields be $y_1, y_2, y_3, y_4$. Let $b_1, b_2$ be the block effects. Then the observational equations are

$$E(y_1) = t_1 + b_1,$$

$$E(y_2) = t_2 + b_1,$$

$$E(y_3) = t_3 + b_2,$$

$$E(y_4) = t_4 + b_2.$$

Suppose we want to estimate the contrast

$$t_1 - t_2 + t_3 - t_4.$$

Then $y_1 - y_2 + y_3 - y_4$ will be an estimate and is independent of the block effects.

Next, suppose we want to estimate the contrast

$$t_1 + t_2 - t_3 - t_4.$$

The estimate will be

$$y_1 + y_2 - y_3 - y_4 - 2(b_1 - b_2)$$

which is not independent of the block effects. In such a case the treatment contrast is said to be confounded with blocks.

A design in which some treatment effects are confounded with blocks, i.e., some treatment effects cannot be estimated independent of block effects is called a confounded design.

Consider the simplest case of a $2^2$ factorial experiment. Let us arrange the 4 treatment combinations as below:

$$B_1 \quad \underset{1}{\overset{y_1}{\boxed{1}}} \; \underset{}{\overset{y_2}{\boxed{d_1 d_2}}} \quad -b_1, \qquad B_2 \quad \overset{y_3}{\boxed{d_1}} \; \overset{y_4}{\boxed{d_2}} \quad -b_2.$$

Let the treatment effects be $t_1, t_2, t_3, t_4$ and yields be $y_1, y_2, y_3, y_4$. Let r be the dividing factor. Then

$$\begin{aligned} D_1 &= r[d_1 - d_2 + d_1 d_2 - 1] \\ &= r[y_3 - y_4 + y_2 - y_1] \\ &= r[t_2 + b_2 - t_3 - b_2 + t_4 + b_1 - t_1 - b_1] \\ &= r[t_2 - t_3 + t_4 - t_1]. \end{aligned}$$

Thus the effect $D_1$ can be estimated independent of block parameters, i.e., apart from the dividing factor, the treatment contrast $t_2 - t_3 + t_4 - t_1$ can be estimated independent of block effects.

$$\begin{aligned} D_2 &= r[d_2 + d_1 d_2 - d_1 - 1] \\ &= r[y_4 + y_2 - y_3 - y_1] \\ &= r[t_3 + t_4 - t_2 - t_1]. \end{aligned}$$

Therefore $D_2$, or equivalently, the treatment contrast $t_3 + t_4 - t_2 - t_1$ can also be estimated independent of the block parameters.

Thus $D_1$ and $D_2$ are not confounded with blocks. Now consider $D_1 D_2$.

$$\begin{aligned} D_1 D_2 &= r[d_1 d_2 + 1 - d_1 - d_2] \\ &= r[y_2 + y_1 - y_3 - y_4] \\ &= r[t_1 + t_2 - t_3 - t_4 + 2(b_1 - b_2)]. \end{aligned}$$

Thus the effect $D_1 D_2$ is confounded with the blocks.

**An Inportant Observation**

If we multiply the elements (effects) in the first block by an element in the same block we will get the same block. If $B_1$ is multiplied by a treatment that does not appear in $B_1$, $B_2$ is obtained. If we multiply $B_2$ by an element in $B_2$, then we get $B_1$.

**Example 7.3**

Consider a feeding experiment in which we want to try two diets (on lambs) at 2 levels, i.e., a $2^2$ factorial experiment in which the experimental plot is a lamb. So we have to get 4 lambs homogeneous in all respects, which is not easy. But twins may be available. So we apply a confounded design with a twin pair of lambs as blocks. Let $b_1, b_2$ be the block effects and let the observations be weights. If we arrange the treatment combinations in the form

$B_1$: | 1 | $d_1d_2$ | $-b_1$; $B_2$: | $d_1$ | $d_2$ | $-b_2$,

then the effect $D_1D_2$ will be confounded with the blocks. We are not interested in the interaction $D_1D_2$, but only in the main effects $D_1$ and $D_2$. So we allow $D_1, D_2$ to be confounded.

## 7.4 $2^3$Factorial Experiment

If it is impossible to get a homogeneous block of 8 plots, but is possible to get 2 homogeneous blocks of 4 plots each, we try the 8 treatment combinations on two such blocks. Let the factors be N, P, K.

$B_1$: | 1 | np | pk | nk | ; $B_2$: | n | p | k | npk |

Consider the effect NPK. In $B_1$, all treatment combinations will have an even number of elements common to NPK. (So all – ve). In $B_2$ all treatment combinations will have an odd number of elements common to NPK ( So all + ve).

Here the effect NPK will be confounded with the blocks. Since all the effects of the first block are – ve and of second are + ve, the block effects do not get cancelled, but get added up.

$$N = (.)[-1 + np - pk + nk + n - k - p + npk],$$

$$P = (.)[-1 + np + pk - nk - n - k + p + npk],$$

$$K = (.)[-1 - np + pk + nk - n + k - p + npk].$$

$(.) =$ Dividing factor.

In each block N, P, K are contrasts. So N, P, K will not be confounded with the blocks. Similarly, NP, PK and NK also.

If we multiply the treatment combinations in the 1-st block by an element in the same block, we get the same block. The $2^3$ elements form a group. But if we consider the 4 elements in $B_1$, they form a group (with self-inverses) w.r.t the operation defined. This group was called by Fisher the intra-block subgroup.

In any confounded design, the treatment combinations in the block containing the unit element will form a group which is called the intra-block subgroup.

Again, $B_1$ multiplied by an element in $B_2$ gives $B_2$. Therefore the treatment combinations in $B_2$ can be obtained by multiplying the treatment combinations in the intra-block subgroup by an element which does not occur in that block. Similarly, $B_2$ multiplied by an element in the same block gives $B_1$, the intra-block subgroup.

If an effect is not confounded with the blocks, it will be a contrast w.r.t. each and every block. In the above case, N,P,K, NP, PK, NK will be contrasts w.r.t. the blocks.

In the above $2^3$ design, the blocks $B_1$ and $B_2$ are not connected. So there are two connected portions in this design, viz. the two blocks. Similarly, a $2^n$ confounded factorial design will be a connected design.

In a connected design with m connected portions the d. f. of the total treatment S.S. is $v - m$ where v is the total number of treatments. Hence in the above example $(v = 8, \ m = 2)$ the treatment S.S. has 6 d. f. which can be split into 1 d. f. each for the S.S. due to N, P, K, NP, PK, NK, which are all estimable.

Consider now r replicates. Then there will be r pairs of blocks.

| | | | | | | | | | | | | | | |
|---|---|---|---|---|---|---|---|---|---|---|---|---|---|---|
| $B_{11}$ | 1 | np | pk | nk | ; $B_{12}$: | | | | | …; $B_{1r}$: | | | | |
| $B_{21}$ | 1 | np | pk | nk | ; $B_{22}$: | | | | | …; $B_{2r}$: | | | | |

If we try the treatment combinations $1, np, pk, nk$ in the 1-st set of blocks then NPK will be even with the 1-st set of blocks. If we apply the treatment combinations $n, p, k, npk$ in the 2-nd set of blocks then NPK will be odd with the 2-nd set of blocks. Since randomization reduces error, we randomize the treatment combinations in a block. Thus we get a $2^3$ confounded factorial experiment with replicates where the effect NPK is confounded with the blocks.

Let y denote an observation in a plot. Then the total variation is

$$\sum y^2 - C \text{ with } (8r - 1) \text{ d.f.}$$

where C is the correction factor. If $B_{ij}$ denotes the block sum in the j-th block ($j = 1,2, \dots, r$) in the i-th set ($i = 1,2$), then the block S.S. is

$$\sum_{i,j} \frac{B_{ij}^2}{4} - C \text{ with } (2r - 1) \text{ d.f.}$$

The d.f. in a replicate for the treatment S.S. is b. So the analysis of the experiment will be as follows:

| Source | S.S. | d.f. |
|---|---|---|
| Total S.S. | $\sum y^2 - C$ | $8r - 1$ |
| Block S.S. | $\sum_{i,j} \frac{B_{ij}^2}{4} - C$ | $2r - 1$ |
| Treatment S.S. (unconfounded) | From Yates' table | 6 |
| Error | (Subtract) | $6(r - 1)$ |

Further analysis will be to divide the treatment S.S. into the S.S. due to the various treatments and that is the advantage of the factorial experiment.

Let $t_1$ be the sum of the r observations corresponding to 1. Similarly, for $t_2, t_3, \dots$

| (1) | (2) | (3) | (4) | (5) |
|---|---|---|---|---|
| 1 | $t_1$ | | | $S_1$ |
| n | $t_2$ | | | $S_2 - N$ |
| p | $t_3$ | | | $S_3 - P$ |
| np | $t_4$ | | | $S_4 - NP$ |
| k | $t_5$ | | | $\vdots$ |
| nk | $t_6$ | | | $\vdots$ |
| pk | $t_7$ | | | $\vdots$ |
| npk | $t_8$ | | | $S_8 - NPK$ |

Apply Yates' process and get columns (3), (4) and (5).

$$\text{S.S. due to } N = \frac{S_2^2}{8r} \text{ with 1 d.f.}$$

$$\text{S.S. due to } P = \frac{S_3^2}{8r} \text{ with 1 d.f.}$$

$$\cdots \quad \cdots \quad \cdots \quad \cdots \quad \cdots \quad \cdots \quad \cdots$$

$$\text{S.S. due to } PK = \frac{S_7^2}{8r} \text{ with 1 d.f.}$$

Therefore

$$\text{Total treatment S.S. (unconfounded)} = \frac{1}{8r}(S_2^2 + S_3^2 + S_4^2 + S_5^2 + S_6^2 + S_7^2) \text{ with 6 d.f.}$$

This is the treatment S.S. in the analysis of variance table. Here there will be no error S.S.

**Inter-Block Information**

Since NPK is confounded with the blocks we will be getting some information regarding the effect NPK from the block S.S. For that we form the following table.

| Replicate / Sign of NPK | 1 | 2 | 3 | ⋯ | r | Sum |
|---|---|---|---|---|---|---|
| + | $B_{21}$ | $B_{22}$ | $B_{23}$ | ⋯ | $B_{2r}$ | $T_1$ |
| − | $B_{11}$ | $B_{12}$ | $B_{13}$ | ⋯ | $B_{1r}$ | $T_2$ |
| Replicate sums | $R_1$ | $R_2$ | $R_3$ | ⋯ | $R_r$ | G |

The individual observations are the block sums. This is an ordinary 2-way classification where each individual observation corresponds to a block total. Therefore the total S.S. (here the block S.S.) can be divided into the S.S. due to blocks (here the blocks are NPK + and NPK -) and the S.S. due to replicates.

| Source | S. S. | d. f. | M. S. S |
|---|---|---|---|
| S. S. due to block | $\sum_{i,j} \frac{B_{ij}^2}{4} - C$ | $2r - 1$ | |
| S. S. due to replicates | $\sum_{j} \frac{R_j^2}{8} - C$ | $r - 1$ | |
| Effect of NPK | (From Yates' process) | 1 | |
| Residual | (Subtract) | $r - 1$ | $V_R$ |

$$\left(\text{Here } C = \frac{G^2}{8r} \text{ is the correction factor}\right).$$

From this table we get some information regarding the effect NPK. This is called recovery of inter-block information. Further if we denote the mean residual S.S. by $V_R$, then we can test the effect NPK by the F-ratio

$$F(1, r-1) = \frac{(\text{S. S. due to NPK})/1}{V_R}.$$

This will not be an accurate test. The complete analysis is given below:

| Source | S.S. | d.f. | M.S.S |
|---|---|---|---|
| Total | $\sum y^2 - C$ | $8r - 1$ | |
| Blocks | $\sum_{i,j} \frac{B_{ij}^2}{4} - C$ | $2r - 1$ | |
| Error | (Subtract) | $6(r - 1)$ | $V_E$ |
| Treatments (unconfounded) | (Yates' table) | 6 | |

| | S.S. | d.f. |
|---|---|---|
| N | ⋮ | 1 |
| P | ⋮ | 1 |
| K | ⋮ | 1 |
| ⋮ | ⋮ | ⋮ |
| NPK | ⋮ | 1 |

$V_E$ can be used for blocks and treatments (unconfounded). $V_R$ can be used for replicates and effect NPK.

Again, let us consider a $2^3$ factorial experiment where 4 blocks are necessary for a single replicate. Let the factors be A, B, C. There are 8 treatment combinations with 2 treatments in a block.

| | | | AB | AC | BC |
|---|---|---|---|---|---|
| I | a | bc | odd | odd | even |
| II | 1 | abc | even | even | even |
| III | b | ac | odd | even | odd |
| IV | c | ab | even | odd | odd |

Consider the effect AB. Then all treatment combinations in Block I have an odd number of elements in common with AB. So the first block is odd as far as the effect AB is concerned. Similarly, second block is even, 3-rd block is odd and 4-th block is even w.r.t. AB. The effect AB can be estimated only if it is a contrast w.r.t. each and every block.

$$AB = \frac{-a - bc}{-2b_1} + \frac{1 + abc}{2b_2} + \frac{-b - ac}{-2b_3} + \frac{c + ab}{2b_4}.$$

So the effect AB is confounded with the blocks.

Similarly AC and BC are also confounded with blocks. Thus all first order interactions are confounded with blocks. But the main effects A, B, C are not confounded with blocks.

$$A = a - bc - 1 + abc - b + ac - c + ab$$
$$= a + ac + ab + abc - 1 - b - c - bc.$$
$$B = b + bc + abc + ab - c - a - 1 - ac.$$

Consider now the 2-nd order interaction ABC.

$$ABC = abc - ab - bc - ca + a + b + c - 1.$$

It is not confounded with the blocks.

Thus the main effects and the 2-nd order interaction are not confounded with the blocks. So if an experimenter is not interested in the 1-st order interactions and if 4 blocks are necessary for a replicate then the above design is suitable.

**Important Observations**

(i) We have seen that when there are 2 blocks in a replicate only one effect is confounded. In the above case where here are 4 blocks in a replicate, there are 3 effects confounded. In general, if there are m blocks in a replicate $(m - 1)$ effects will be confounded.

(ii) In a confounded design all the confounded effects will be even w.r.t. the intra-block subgroup.

**Theorem 7.1**

If X and Y are confounded effects (with the blocks) then their generalized interaction XY also will be confounded with the blocks.

**Proof.**

Consider a $2^n$ design where the effects X and Y are confounded with the blocks. So if we consider a particular block it will be either even w.r.t. X or odd w.r.t. Y and it will be either even w.r.t. Y or odd w.r.t. X. If the block is odd w.r.t. both X and Y, then it will be even w.r.t. XY. If it is even w.r.t. both X and Y, then it will be even w.r.t. XY. If the blocks is even (odd) w.r.t. X and odd (even) w.r.t. Y, then it will be odd w.r.t. XY.

Also, in a block all the treatments should be either even or odd and not some even and some odd. So the effect XY is also confounded with the blocks.

The confounded effects will be even w.r.t. the intra-block subgroup. We noted earlier that if a number of effects are even w.r.t. the intra-block subgroup then those effects will be confounded with the blocks. So if we want to confound a certain number of effects, then we select the intra-block subgroup in such a way that the effects have even number of elements common to the treatment combinations in the intra-block subgroup. In the $2^3$ experiment with 4 blocks in a replicate we can confound 3 effects. Suppose we want to confound the effects AB, BC, AC. The intra-block subgroup must contain the unit element 1. Hence we need select only one more treatment combination which will have an even number of elements common to these effects. In this case the intra-block subgroup will be

| 1 | abc |
|---|---|

.

Multiplying by a, the second block is

| a | bc |
|---|---|

.

Multiplying by b, the 3-rd block is

| b | ac |
|---|---|

.

Multiplying by c, the 4-th block is

| c | ab |
|---|---|

.

So the complete design will be

| 1 | abc |
|---|---|
| a | bc |

| b | ac |
|---|---|
| c | ab |

It is enough to confound only AB and BC, for then AC will be automatically confounded since it is the generalized interaction of AB, BC. ■

**Case of a $2^4$ Experiment**

Let there be 4 blocks in a replicate. These effects can be confounded. Let the effects AB, CD be confounded. The ABCD will also be confounded. The design will be

| 1 | ab | cd | abcd |
|---|---|---|---|
| a | b | acd | bcd |
| c | abc | d | abd |
| bc | ac | bd | ad |

We can confound certain effects in all the replicates. Then we will get no information regarding these effects? This is the case of total confounding.

If some effects are confounded in some replicates and not confounded in some other replicates then we can get information regarding all the effects from one or the other of the replicates. This is the case of partial confounding. Partial confounding is useful when the experimenter wants some information regarding all the effects.

Let there be r replicates in s of which an effect is confounded. Then we say that the effect is estimated with a precision $\frac{r-s}{r}$. A design in which all the effects are estimated with the same precision is called a balanced design. If it is also a confounded design, we call it a balanced partially confounded design.

**Example 7.4**

$2^2$ factorial designs with 3 replicates.

| I | II | III | IV |
|---|---|---|---|
| 1 \| ab | 1 \| a | 1 \| b | 1 \| ab |
| a \| b | b \| ab | a \| ab | a \| b |
| A, B confounded (Therefore AB also) | A is confounded. | B is confounded. | AB confounded. |

AB is estimated with a precision 2/3. Similarly, the effects A and B are also. This is a balanced partially confounded design.

If there is a 4-th replicate with AB confounded then

$$\text{Precision of AB} = \frac{2}{4} = \frac{1}{2},$$

$$\text{Precision of A} = \frac{3}{4},$$

$$\text{Precision of AB} = \frac{3}{4}.$$

So the design is not balanced.

**Analysis of a $2^3$ Factorial Experiment**

Let there be 4 blocks in a replicate with 3 effects totally confounded in r replicates. There are 8 plots in a replicate and there are r replicates. Also there are 8 treatment combinations and $m = 4$ connected portions in a replicate. Each block is a connected portion. The division of the d. f. is as follows:

| Total | $8r - 1$ |
|---|---|
| Blocks | $4r - 1$ |
| Treatments (unconfounded) | $v - m = 8 - 4 = 4$ |

Each of the 4 unconfounded effects will have one d. f. and so

$$\text{error d.f.} = (8r-1) - (4r-1) - 4 = 4(r-1).$$

Let the independent confounded effects be $E_1'$ and $E_2'$, then $E_1'E_2'$ will also be confounded. Then we can divide the 4 sets of blocks such that the 1-st set are + + w.r.t. the independent confounded effects, the 2-nd set are + -, the 3-rd are - + and the 4-th are - -.

| Signs of the independent confounded effects w.r.t. the blocks | Replicate | | | | |
|---|---|---|---|---|---|
| | 1 | 2 | ... | r | Sum |
| + +<br>+ −<br>− +<br>− − | Block Sums | | | | $T_1$<br>$T_2$<br>$T_3$<br>$T_4$ |
| Sum | $R_1$ | $R_2$ | | $R_r$ | G |

The complete block analysis will be:

| Variation due to | S. S. | d. f. | M. S. S. | F |
|---|---|---|---|---|
| Blocks | $\sum_{i,j} \frac{B_{ij}^2}{r} - C$ | $4r-1$ | | |
| Replicates | $\sum_i \frac{R_i^2}{8} - C$ | $r-1$ | $V_R'$ | $\frac{V_R'}{V_R}$ |
| Treatments (confounded) | $\sum_i \frac{T_i^2}{2r} - C$ | 3 | $V_T'$ | $\frac{V_T'}{V_R}$ |
| Residual | (sub) | $3(r-1)$ | $V_R$ | |

$\frac{V_T'}{V_R}$ will provide test for information regarding the confounded effects. The confounded effects as such cannot be tested since they are not estimable.

Complete Treatment Analysis – Yates' table

| (1) | (2) | (3) | (4) | (5) | S. S. |
|---|---|---|---|---|---|
| 1 | $t_1$ | $S_1$ | $S_1'$ | $S_1''$ | |
| a | $t_2$ | $S_2$ | $S_2'$ | $S_2''$ | |
| b | $t_3$ | $S_3$ | $S_3'$ | $S_3''$ | $\frac{(S_3'')^2}{8r}$ |
| ab | $t_4$ | $S_4$ | $S_4'$ | $S_4''$ | $\frac{(S_4'')^2}{8r}$ |
| c | $t_5$ | $D_1$ | $D_1'$ | $D_1''$ | |
| ac | $t_6$ | $D_2$ | $D_2'$ | $D_2''$ | $\frac{(D_2'')^2}{8r}$ |
| bc | $t_7$ | $D_3$ | $D_3'$ | $D_3''$ | $\frac{(D_3'')^2}{8r}$ |
| abc | $t_8$ | $D_4$ | $D_4'$ | $D_4''$ | |

| Variation due to | S. S. | d. f. | M. S. S. | F |
|---|---|---|---|---|
| Effect B | $\frac{(S_3'')^2}{8r}$ | 1 | $V_1$ | $\frac{V_1}{V_E}$ |
| Effect AB | $\frac{(S_4'')^2}{8r}$ | 1 | $V_2$ | $\frac{V_2}{V_E}$ |
| Effect C | $\frac{(D_2'')^2}{8r}$ | 1 | $V_3$ | $\frac{V_3}{V_E}$ |
| Effect AC | $\frac{(D_3'')^2}{8r}$ | 1 | $V_4$ | $\frac{V_4}{V_E}$ |
| Total | Sum | 4 | | |

Here $V_E$ is the error S.S. in the general analysis and $\frac{V_i}{V_E}, i = 1,2,3,4$ will all be $F(1,4(r-1))$.

The above is the analysis in the case of total confounding.

**Analysis in the Case of Partial Confounding**

Illustration in the case of a $2^2$ factorial experiment:

I

$b_1$: | 1 | ab | $-B_{11}$ $\qquad$ $b_3$: | 1 | a | $-B_{12}$

$b_2$: | a | b | $-B_{21}$ $\qquad$ $b_4$: | b | ab | $-B_{22}$

AB confounded $\qquad$ A confounded

$b_5$: | 1 | b | $-B_{13}$

$b_6$: | a | ab | $-B_{23}$

B confounded

Each of the confounded effects is estimated with a precision 2/3. Therefore this is a balanced partially confounded design. Yates' process for the treatment analysis is as follows:

| (1) | (2) | (3) | (4) | Adjusted treatment Totals | |
|---|---|---|---|---|---|
| | | | | A + | A - |
| 1 | $t_1$ | $S_1$ | $S'_1$ | | |
| a | $t_2$ | $S_2$ | $S'_2$ | $S'_2 - (B_{12} - B_{22})$ | |
| b | $t_3$ | $D_1$ | $D'_1$ | $D'_1 - (B_{13} - B_{23})$ | |
| ab | $t_4$ | $D_2$ | $D'_2$ | $D'_2 - (B_{11} - B_{21})$ | |

But here $\frac{(S'_2)^2}{2^2 \times 3}$ is not the S.S. due to A. We have to divide it by the effective number of replicates which is $r - s$ if the effect is confounded in s out of r replicates. In the case of a $2^n$ factorial experiment with r replicates the dividing factor will be $2^n(r - s)$. In our example the dividing factor is $2^2(3 - 1) = 8$. Again,

$$\begin{aligned} S'_2 &= D_1 + D_2 \\ &= t_2 - t_1 + t_4 - t_3 \\ &= (2b_3 - 2b_4) + (\text{the corresponding treatment sums}) \end{aligned}$$

Therefore it is not possible to get the actual value of $S'_2$. We make an adjustment for $S'_2$ and take instead $S'_2 - (B_{12} - B_{22})$ as the new $S'_2$.

Similar adjustments are made in $D_1'$ and $D_2'$. Thus

$$\text{S.S. due to A} = \frac{[S_2' - (B_{12} - B_{22})]^2}{2^2(3-1)} \quad \text{with 1 d.f.}$$

$$\text{S.S. due to B} = \frac{[D_1' - (B_{13} - B_{23})]^2}{2^2(3-1)} \quad \text{with 1 d.f.}$$

$$\text{S.S. due to AB} = \frac{[D_2' - (B_{11} - B_{21})]^2}{2^2(3-1)} \quad \text{with 1 d.f.}$$

The general analysis will be

| | d.f. | M.S.S. |
|---|---|---|
| Total | 11 | |
| Blocks | 5 | |
| Treatments | 3 | |
| Error | 3 | $V_E$ |

Treatment S.S. is obtained from Yates' process.

In a $2^2$ factorial experiment which is not balanced, the only changes in the analysis will be the adjustments to be made in the dividing factor and the adjusted treatment effects.

Consider a $2^4$ factorial experiment. What will be the minimum number of replicates required where a replicate contains 2 blocks such that the main effects and the first order interactions are not confounded and the confounded effects are balanced? Let A, B, C, D be the factors with ABC, ABD, ACD, BCD, and ABCD as the confounded effects. Here $\binom{4}{3}$ is the total number of 2-nd order interactions and $\binom{4}{4} = 1$ is the total number of 3-rd order interactions. So there are $\binom{4}{3} + \binom{4}{4} = 5$ interactions which can be confounded in a replicate, and the minimum number of replicates required is 5.

Again, consider a $2^5$ factorial experiment with 4 blocks per replicate. In a single block there will be 8 treatment combinations. Let us get an arrangement such that ABC, and CDE are confounded with the blocks. The intra-block subgroup will be

| 1 | ab | de | abde | ace | bce | acd | bcd |
|---|---|---|---|---|---|---|---|

In this ab, de, ace are the only independent treatment combinations. So if we are given these three independent treatment combinations, then we can get the intra-block subgroup and so all other blocks. So if we are given 2 independent effects or 3 independent treatment combinations the system is completely specified. Note that

$$2^5 = 2^2 \times 2^3$$
$$= (\text{No. of blocks per replicate}) \times (\text{No. of units per block}).$$

The index of the first factor on the right gives the number of independent effects that can be confounded and the index of the second factor gives the number of independent treatment combinations in the intra-block subgroup.

## Generalization

Consider a $2^n$ factorial experiment. Let

$$2^n = 2^m \times 2^{n-m},$$

so there will be $2^m$ blocks per replicate and $2^{n-m}$ will be the total number of plots per block. If we are given m independent effects which are confounded with the blocks then the system is completely specified. Or, if we are given $(n - m)$ independent treatment combinations, then also the system is completely specified.

If the number of blocks per replicate is fixed, can we get an arrangement such that interactions up to a specified order are not confounded with the blocks? Problem was posed by Fisher who constructed a design such that interactions up to order 2 were confounded. R. C. Bose has generalized this construction.

## Fisher's Construction

Under the assumption that the number of factors is less than the number of treatments per block, a $2^n$ factorial design can be constructed such that the main effects and 1-st order interactions are not confounded.

Let there be $2^n$ blocks of $2^{n-m}$ plots each. Let the number of factors be $2^r - 1$ where $r = n - m$. Let $\alpha_1, \alpha_2, \ldots, \alpha_r$ be r elements. Then we form a set S of combinations of these elements in the form:

taking none at a time, one at a time, two at a time, and so on. Then $2^r$ combinations are possible with the r elements $\alpha_1, \alpha_2, \dots, \alpha_r$ together with 1. Let the factors be $A_1, A_2, \dots, A_{2^r-1}$ and let

$$\begin{aligned} &\alpha_1 \to A_1 \\ &\alpha_2 \to A_2 \\ &\alpha_1\alpha_2 \to A_3 \\ &\alpha_3 \to A_4 \\ &\alpha_1\alpha_3 \to A_5 \\ &\alpha_2\alpha_3 \to A_6 \\ &\alpha_1\alpha_2\alpha_3 \to A_7 \end{aligned}$$

and so on. Thus we can establish a correspondence between the effects and the elements of S.

Let us specify a particular combination $\alpha_i\alpha_j\alpha_k$. Consider those letter combinations which are having an odd number of letters common with this specified letter combination. For instance consider $\alpha_1\alpha_2$ as the particular combination in a $2^3$ design with $2^3 - 1$ factors. Then $\alpha_1, \alpha_2, \alpha_1\alpha_3, \alpha_2\alpha_3$ all have an odd number of letters common with $\alpha_1\alpha_2$. Consider the factors which correspond to $\alpha_1, \alpha_2, \alpha_1\alpha_3, \alpha_2\alpha_3$. Take the product of the levels of these factors. Let it correspond to $\alpha_1\alpha_2$. Then

$$\alpha_1\alpha_2 \to a_1a_2a_5a_6.$$

If we specify $\alpha_3$, then $\alpha_1\alpha_3, \alpha_2\alpha_3, \alpha_1\alpha_2\alpha_3$ may be taken and $\alpha_3 \to a_4a_5a_6a_7$. Similarly, $\alpha_2\alpha_3 \to a_2a_3a_4a_5$ and $\alpha_1\alpha_3 \to a_1a_3a_4a_6$. This correspondence is one-one. Let

$$x \to t, \; x' \to t'.$$

Let $x = \alpha_1\alpha_2$, and $x' = \alpha_2\alpha_3$. Then $t = a_1a_2a_5a_6$, and $t' = a_2a_3a_4a_5$. So

$$xx' = \alpha_1\alpha_3 \to a_1a_3a_4a_6 = tt'.$$

Thus the correspondence is an isomorphism.

Consider the letter combinations taking 1 also. The set will form a group, and by the isomorphism we have established, the treatment combinations together with 1 will form a group, viz. the intra-block subgroup. Let

$$\begin{aligned}
\alpha_1 &\rightarrow A_1(a_1a_3a_5a_7)\\
\alpha_2 &\rightarrow A_2(a_2a_3a_6a_7)\\
\alpha_1\alpha_2 &\rightarrow A_3(a_1a_2a_5a_6)\\
\alpha_3 &\rightarrow A_4(a_4a_5a_6a_7)\\
\alpha_1\alpha_3 &\rightarrow A_5(a_1a_3a_4a_6)\\
\alpha_2\alpha_3 &\rightarrow A_6(a_2a_3a_4a_5)\\
\alpha_1\alpha_2\alpha_3 &\rightarrow A_7(a_1a_2a_4a_7)
\end{aligned}$$

These treatment combinations together with 1 form the intra-block subgroup. By the same procedure, if the number of plots is $2^r$ and the number of factors is $2^r - 1$, we will be able to get the intra-block subgroup.

Now we can prove that the main effects and 1-st order interactions are not confounded with the blocks. Consider the effect $X = A_iA_j \dots A_k$. Let us call $\alpha_1, \alpha_2, \alpha_1\alpha_2, \dots$ the group elements. Corresponding to the interaction X we will get a group element 1 or different from 1.

**Case (i)**

Let the group element corresponding to X be 1. Then, corresponding to 1 we will be getting 1 in the intra-block subgroup and 1 is even w.r.t. X. Therefore X will be even w.r.t. the intra-block subgroup, and so will be confounded with the intra-block subgroup.

**Case (ii)**

Let the element corresponding to X in this group be $\alpha(\neq 1)$. Then $\alpha$ will be present in an odd number of elements in the group corresponding to the A's in X. Therefore $\alpha$ should be present in an odd number of times in X and X is not confounded with the blocks.

**(i) Main Effects**

Let y be a main effect so that it contains only one factor and correspondingly there is only one group element, which is different from1. So y is not confounded with the blocks.

**(ii) First order interactions**

Let Z be a first order interaction. There will be two factors of the type $A_i, A_j$ in Z. Therefore there will be 2 group elements corresponding to $A_i, A_j$ etc.

**Illustration**

Consider 3 factors $A_1, A_2, A_3$. Let

$$\alpha_1 \to A_1, \quad \alpha_2 \to A_2, \quad \alpha_1\alpha_2 \to A_3.$$

$A_1, A_2, A_3$ are not confounded with the blocks. Consider the interaction $A_1A_2$. The corresponding group element is $\alpha_1\alpha_2 \neq 1$. Therefore $A_1A_2$ is not confounded with the blocks. Similarly, the group element corresponding to $A_1A_3$ is $\alpha_1\alpha_2\alpha_3 = \alpha_3 \neq 1$. Hence $A_1A_3$ is not confounded with the blocks. Similarly, $A_2A_3$ is also not confounded with the blocks.

Consider the group product corresponding to $A_1A_2A_3$. It is $\alpha_1\alpha_2 \cdot \alpha_1\alpha_2 = 1$. So $A_1A_2A_3$ is confounded with the blocks.

The elements in the intra-block subgroup corresponding to $\alpha_1, \alpha_2, \alpha_1\alpha_2$ are $a_1a_3, a_2a_3, a_1a_2$. So the intra-block subgroup is

$$\boxed{1 \;\; a_1a_3 \;\; a_2a_3 \;\; a_1a_2}$$

**7.5 Fractional Replications**

In a design of fractional replications we will be studying the effects of only a fraction of the total treatment combinations.

In a $2^2$ factorial design with effects A, B, the effect A can be represented in the form

$$A = ab + a - b - 1\,.$$

Similarly,

$$B = ab + b - a - 1$$

$$AB = ab + 1 - a - b,$$

and

$$A + AB = (.)2(ab - b), \;\; B + AB = (.)2(ab - a)$$

where $(.)$ is the dividing factor.

The mean M can be written as $ab + 1 + a + b$. Therefore

$$M + AB = (.)2(ab + 1).$$

Therefore the effect A+AB can be estimated by 2 treatment combinations ab and b. So if we use only 2 treatment combinations ab and b we will be able to study the effect A+AB. If we know that the interaction AB is insignificant we will be estimating the main effect A. Similarly for B+AB.

Consider a $2^3$ experiment with factors N, P, K.

| | npk | p | k | n | np | pk | nk | 1 |
|---|---|---|---|---|---|---|---|---|
| N | + | − | − | + | + | − | + | − |
| PK | + | − | − | + | − | + | − | + |
| P | + | + | − | − | + | + | − | − |
| NK | + | + | − | − | − | − | + | + |
| K | + | − | + | − | − | + | + | − |
| NP | + | − | + | − | + | − | − | + |
| NPK | + | + | + | + | − | − | − | − |
| M | + | + | + | + | + | + | + | + |

$$N + PK = (.)2(npk - p - k + n)$$

$$P + NK = (.)2(npk + p - k - n)$$

$$K + NP = (.)2(npk - p + k - n)$$

$$NPK + M = (.)2(npk + p + k + m).$$

So with the 4 treatment combinations $npk, n, p, k$ we will be able to estimate the effects $N + PK, P + NK, K + NP,$ and $M + NPK$. All these can be estimated by considering only a fraction of the total treatment combinations. We will not be able to estimate the main effects or the 1-st or 2-nd order interactions separately.

By considering the fraction $np, pk, nk, 1$ of the total treatment combinations we will be able to estimate the effect $N - PK, P - NK, K - NP$ and $NPK - M$.

Consider the first case. The effects $N + PK, P + NK, K +$

$NP, M + NPK$ can be studied by the fraction $npk, n, p, k$ of the treatment combinations. N cannot be estimated independent of PK. Thus N and PK are together estimable, but not separately. In such cases each is called the alias of the other. Thus

$$\text{N is the alias of PK,}$$

$$\text{PK is the alias of N,}$$

$$\text{P is the alias of NK,}$$

$$\text{NK is the alias of P, etc.}$$

The relationship "NPK is the alias of M" is written symbolically as $NPK = I$, and this is called the defining relationship. It will define the whole design. For example, multiplying by P, we get $NK = P$, i.e., NK is the alias of P. Similarly for the others.

We need consider only those treatment combinations which will be $+$ vc w.r.t. NPK, i.e., the treatment combinations $npk, n, p, k$. With these treatment combinations we can construct the whole design.

Consider the second fraction of the treatment combinations. Here $NPK - M$ is estimable, and the defining relationship is $NPK = -I$. Multiplying by P,

$$NK = -P \quad \text{etc.}$$

If we use 4 treatment combinations out of 8, i.e., half the total number of treatment combinations, we can estimate 2 effects together. This is a general property. In general, if we use $\frac{1}{n}$-th of the total treatment combinations we can estimate n effects together. This is called a $\frac{1}{n}$-th replication.

## 7.6 Confounding in Latin Squares

Consider a $2^3$ factorial experiment with factors N, P, K. The treatment combinations are $1, n, p, k, np, pk, nk, npk$.

Consider a $4 \times 4$ Latin square. Suppose that there are 2 blocks in a replicate and suppose we want to confound the effect $npk$. Then the blocks in one replicate will be

| 1 | np | pk | nk |
|---|---|---|---|
| n | p | npk | k |

If there are 2 replicates, then the 4 blocks will be

| 1 | np | pk | nk |
|---|---|---|---|
| n | p | npk | k |
| 1 | np | pk | nk |
| n | p | npk | k |

These can be considered to be the 4 rows of a Latin square with treatments arranged properly. Here the effect NPK is confounded with the rows and PK is confounded with the columns. This may therefore be called a confounded Latin square arrangement.

| 1 | np | pk | nk |
|---|---|---|---|
| n | p | npk | k |
| pk | nk | 1 | np |
| npk | k | n | p |

(Here also we achieve a 2-way elimination of heterogeneity.)

**ANALYSIS**

Consider 2 Latin squares as above.

| Variation due to | d.f. |
|---|---|
| Squares | 2 |
| Rows within squares | $3 + 3 = 6$ |
| Columns within squares | $3 + 3 = 6$ |
| Treatments | 5 |
| Error | 13 |
| Total | 32 |

Further analysis of treatment S.S. is obtained from Yates' table.

The treatment S.S. can be divided into the S.S. due to the 5 estimable effects and each will have one d.f.

## 7.7 Partial Confounding

Suppose we confound NPK and PK in the 1-st square and NPK and NP in the second. Then we can get information regarding the NP from the first and regarding the PK from the second.

| I | | | | II | | | |
|---|---|---|---|---|---|---|---|
| 1 | np | pk | nk | 1 | np | pk | nk |
| n | p | npk | k | np | 1 | nk | pk |
| pk | nk | 1 | np | k | npk | p | n |
| npk | k | n | p | npk | k | n | p |

This is a partially confounded Latin square arrangement. Only NPK is completely confounded with the blocks. So there will be a change in the final analysis. 6 treatment effects are estimable. So the treatment S.S. will have 6 d. f. and then the error S.S. will have only 12 d. f. The treatment analysis will be

| Variation due to | d. f. |
|---|---|
| N | 1 |
| P | 1 |
| K | 1 |
| NP | 1 |
| NK | 1 |
| PK | 1 |
| Total | 6 |

There will be adjustments in the Yates process also.

Suppose we want to get information on all the effects. Then in II we do not confound PK and NPK, but some other effects. Suppose we confound NK with the rows and NP with the columns in II. Then on the whole 4 effects are partially confounded.

| II | | | |
|---|---|---|---|
| 1 | nk | p | npk |
| np | pk | n | k |
| k | n | pk | np |
| npk | p | nk | 1 |

Here the d.f. for the treatments will be 7 and for the error 11. There will be adjustments in the Yates' process for the effects NP, PK, NPK and NK. The treatment analysis will be

| Variation due to | d.f. |
|---|---|
| N | 1 |
| P | 1 |
| K | 1 |
| NP | 1 |
| NK | 1 |
| PK | 1 |
| NPK | 1 |
| Total | 7 |

S.S. due to N, P, K can be obtained from Yates' process directly. For the others there will be adjustments.

**Factors at 3 Levels**

Consider a $3^2$ factorial design, i.e., 2 factors at 3 levels. Let A, B be the factors and let the levels be

| | | |
|---|---|---|
| $a_0$ | $a_1$ | $a_2$ |
| $b_0$ | $b_1$ | $b_2$ |

There are $3^2 = 9$ treatment combinations, viz.,
$a_0b_0, a_0b_1, a_0b_2, a_1b_0, a_1b_1, a_1b_2, a_2b_0, a_2b_1, a_2b_2$.

| B \ A | 0 | 1 | 2 | Sum |
|---|---|---|---|---|
| 0 | $a_0b_0$ | $a_1b_0$ | $a_2b_0$ | $R_0$ |
| 1 | $a_0b_1$ | $a_1b_1$ | $a_2b_1$ | $R_1$ |
| 2 | $a_0b_2$ | $a_1b_2$ | $a_2b_2$ | $R_2$ |
| Sum | $C_0$ | $C_1$ | $C_2$ | |

A, B may be 2 manures, say A-potassium, B-super phosphate, at levels

$$A - 0 \text{ lbs}, \quad 50 \text{ lbs}, \quad 100 \text{ lbs}.$$

$$B - 0 \text{ lbs}, \quad 20 \text{ lbs}, \quad 40 \text{ lbs}.$$

The difference $C_1 - C_0$ is caused by the level difference in A. Therefore $C_1 - C_0$ is a measure of the level difference in A, similarly $C_2 - C_1$. Hence the average

$$\frac{(C_2 - C_1) + (C_1 - C_0)}{2} = \frac{C_2 - C_0}{2}$$

is a measure of the level difference in A.

$\frac{1}{2}(C_2 - C_0)$ is called the linear effect of A. Similarly, $\frac{1}{2}(R_2 - R_0)$ is the linear effect of B. If $(C_2 - C_1) - (C_1 - C_0) = 0$, the effect of A is uniform. If $(C_2 - C_1) - (C_1 - C_0) > 0$, the effect is increasing and if it is $< 0$ the effect is decreasing.

$\frac{1}{2}[(C_2 - C_1) - (C_1 - C_0)]$ is called the quadratic effect of A. Similarly, $\frac{1}{2}[(R_2 - R_1) - (R_1 - R_0)]$ is called the quadratic effect of B.

Let x denote the levels and $C_0, C_1, C_2$ the effects.

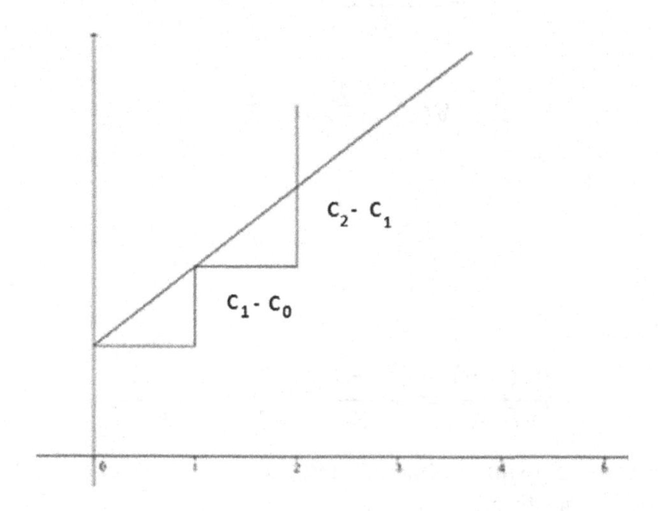

| x | y |
|---|---|
| 0 | $C_0$ |
| 1 | $C_1$ |
| 2 | $C_2$ |

Consider the graph of x and y.

If $C_2$ lies on the line joining $(0, C_0)$ and $(1, C_1)$, then only

$$(C_2 - C_1) - (C_1 - C_0) = 0.$$

The relation $y = f(x)$ is called the response surface.

Let us find the regression coefficient assuming linear relationship.

$$\text{Regression coefficient} = \frac{\text{Cov}(x, y)}{V(x)}.$$

$$\begin{aligned}\text{Cov}(x, y) &= \frac{2xy}{n} - \bar{x}\bar{y} \\ &= \frac{C_1 + 2C_2}{3} - 1 \cdot \frac{C_0 + C_1 + C_2}{3} \\ &= \frac{C_2 - C_0}{3}\end{aligned}$$

and

$$\begin{aligned}V(x) &= \frac{\sum x^2}{n} - \bar{x}^2 \\ &= \frac{5}{3} - 1 \\ &= \frac{2}{3}.\end{aligned}$$

Therefore

$$\frac{\text{Cov}(x, y)}{V(x)} = \frac{C_2 - C_0}{3} \times \frac{3}{2} = \frac{C_2 - C_0}{2}.$$

Thus if the relationship is linear, then the regression coefficient is the same as the linear effect.

Let us assume that y is a quadratic function of x, say $y = ax^2 + bx + c$. If we change the origin to $(-1,0)$, then

| x | y |
|---|---|
| $-1$ | $C_0$ |
| 0 | $C_1$ |
| 1 | $C_2$ |

The least square normal equations are

$$\sum y = a\sum x^2 + b\sum x + nc$$

$$\sum xy = a\sum x^3 + b\sum x^2 + c\sum x$$

$$\sum x^2y = a\sum x^4 + b\sum x^3 + c\sum x^2.$$

After necessary computations, these are

$$C_0 + C_1 + C_2 = 2a + 3c$$

$$C_2 - C_0 = 2b$$

$$C_2 + C_0 = 2a + 2c.$$

Solving, we get

$$a = \frac{C_2 - 2C_1 + C_0}{2}, \quad b = \frac{C_2 - C_0}{2}, \quad c = C_1.$$

The coefficient of $x^2$ is the quadratic effect of A.

Consider levels of A in the form:

| 0 lbs | 30 lbs | 70 lbs |
|---|---|---|
| $C_0 = 85$ lbs | $C_1 = 90$ lbs | $C_2 = 100$ lbs (observations) |

Note that here the level is not uniform. Only when the level is uniform we define the linear and quadratic effects. Here also we can construct a contrast.

$\frac{1}{3}(C_1 - C_0)$ is a measure of the effect due to increase of 10 lbs. Similarly, $\frac{1}{4}(C_2 - C_1)$ is a measure of the increase in effect due to a 10 lbs increase in level. Therefore

$$\frac{C_1 - C_0}{3} + \frac{C_2 - C_1}{4} = \frac{3C_2 + C_1 - 4C_0}{12}$$

will measure the effect of A for an increase of 10 lbs. This is a contrast.

$$\begin{aligned} C_0 &= a_0b_0 + a_0b_1 + a_0b_2 \\ C_1 &= a_1b_0 + a_1b_1 + a_1b_2 \\ C_2 &= a_2b_0 + a_2b_1 + a_2b_2. \end{aligned}$$

In general $l_0C_0 + l_1C_1 + l_2C_2$ such that $l_0 + l_1 + l_2 = 0$ will be a contrast. The linear effect of A apart from the dividing factor is $L_A = C_2 - C_0$. Now,

$$\begin{aligned} C_0 &= a_0(b_0 + b_1 + b_2) \\ C_1 &= a_1(b_0 + b_1 + b_2) \\ C_2 &= a_2(b_0 + b_1 + b_2). \end{aligned}$$

Therefore $L_A = C_2 - C_0 = (a_2 - a_0)(b_0 + b_1 + b_2)$. Similarly $L_B = (b_2 - b_0)\,(a_0 + a_1 + a_2)$. The quadratic effect of A is

$$Q_A = C_2 - 2C_1 + C_0 = (a_2 - 2a_1 + a_0)(b_0 + b_1 + b_2).$$

Similarly,

$$Q_B = (b_2 - 2b_1 + b_0)(a_0 + a_1 + a_2).$$

In general, let $T_0, T_1, T_2$ be the sums corresponding to a particular factor. Then apart from the dividing factor the general mean is given by

$$M = T_0 + T_1 + T_2.$$

The general linear effect is given by

$$L = T_2 - T_0$$

and the general quadratic effect is

$$Q = T_2 - 2T_1 + T_0.$$

Normalization

Let

$$\frac{M}{\sqrt{3}} = \frac{T_0 + T_1 + T_2}{\sqrt{3}},$$

$$\frac{L}{\sqrt{2}} = \frac{T_2 - T_0}{\sqrt{2}},$$

$$\frac{Q}{\sqrt{6}} = \frac{T_2 - 2T_1 + T_0}{\sqrt{6}}.$$

Then

$$E = \left(\frac{M}{\sqrt{3}}, \frac{L}{\sqrt{2}}, \frac{Q}{\sqrt{6}}\right)$$

$$= (T_0 \quad T_1 \quad T_2)\begin{pmatrix} \frac{1}{\sqrt{3}} & -\frac{1}{\sqrt{2}} & \frac{1}{\sqrt{6}} \\ \frac{1}{\sqrt{3}} & 0 & -\frac{2}{\sqrt{6}} \\ \frac{1}{\sqrt{3}} & \frac{1}{\sqrt{2}} & \frac{1}{\sqrt{6}} \end{pmatrix}$$

$$= TX, \text{ say}$$

But $XX^T = I$. So

$$EE^T = TT^T = \sum T_i^2 = T_0^2 + T_1^2 + T_2^2.$$

$$\text{i.e., } \frac{M^2}{3} + \frac{L^2}{2} + \frac{Q^2}{6} = T_0^2 + T_1^2 + T_2^2.$$

Therefore

$$\frac{L^2}{2} + \frac{Q^2}{6} = T_0^2 + T_1^2 + T_2^2 - \frac{M^2}{3},$$

is the total S.S. due to the factor. Thus the total variation due to a factor can be divided into the variation due to the linear effect and the variation due to the quadratic effect. This is the principle behind splitting the S.S. due to a factor.

Here $a_2b_0 - a_0b_0$ is a measure of the linear effect of A at the level $b_0$, and $a_2b_2 - a_0b_2$ is a measure of the linear effect of A at the level $b_2$. Therefore $(a_2b_2 - a_0b_2) - (a_2b_0 - a_0b_0)$ is a measure of the interaction (apart from the dividing factor) between $L_A$ and $L_B$, written $L_AL_B$. Thus

$$L_AL_B = (a_2b_2 - a_0b_2) - (a_2b_0 - a_0b_0)$$
$$= (a_2 - a_0)(b_2 - b_0).$$

Similarly we can get expressions for interactions of the type $L_AQ_B, L_BQ_A$ and $Q_AQ_B$. Therefore we can get 4 types of interactions in a $3^2$ factorial design.

(i) interaction between linear effects

(ii) 2 interactions between linear and quadratic effects

(iii) interaction between the quadratic effects.

The total of 8 d. f. can be divided into one each for the 2 linear effects, one each for the 2 quadratic effects and one each for the 4 interactions.

$$L_AL_B = (a_2 - a_0)(b_2 - b_0)$$

$$L_AQ_B = (a_2 - a_0)(b_2 - 2b_1 + b_0)$$

$$Q_AL_B = (a_2 - 2a_1 + a_0)(b_2 - b_0)$$

$$Q_AQ_B = (a_2 - 2a_1 + a_0)(b_2 - 2b_1 + b_0).$$

Table of coefficients:

| | $a_0b_0$ $T_1$ | $a_0b_1$ $T_2$ | $a_0b_2$ $T_3$ | $a_1b_0$ $T_4$ | $a_1b_1$ $T_5$ | $a_1b_2$ $T_6$ | $a_2b_0$ $T_7$ | $a_2b_1$ $T_8$ | $a_2b_2$ $T_9$ |
|---|---|---|---|---|---|---|---|---|---|
| M | 1 | 1 | 1 | 1 | 1 | 1 | 1 | 1 | 1 |
| $L_A$ | −1 | −1 | −1 | 0 | 0 | 0 | 1 | 1 | 1 |
| $L_B$ | −1 | 0 | 1 | −1 | 0 | 1 | −1 | 0 | 1 |
| $Q_A$ | 1 | 1 | 1 | −2 | −2 | −2 | 1 | 1 | 1 |
| $Q_B$ | 1 | −2 | 1 | 1 | −2 | 1 | 1 | −2 | 1 |
| $L_AL_B$ | 1 | 0 | −1 | 0 | 0 | 0 | −1 | 0 | 1 |
| $L_AQ_B$ | | | | | | | | | |
| $Q_AL_B$ | | | | | | | | | |
| $Q_AQ_B$ | | | | | | | | | |

Let the corresponding observations be $T_1, T_2, \ldots, T_9$. Then

$$\frac{M}{\sqrt{9}} = \frac{T_1 + T_2 + \cdots + T_9}{\sqrt{9}},$$

$$\frac{L_A}{\sqrt{6}} = \frac{-T_1 - T_2 - T_3 + 0 \cdot T_4 + 0 \cdot T_5 + 0 \cdot T_6 + T_7 + T_8 + T_9}{\sqrt{6}}, \text{ etc.}$$

Therefore

$$\left(\frac{M}{\sqrt{9}}, \frac{L_A}{\sqrt{6}}, \ldots, \frac{Q_A Q_B}{\sqrt{36}}\right) = TX$$

where $T = (T_1, T_2, \ldots, T_9)$. i.e., $E = TX$. Therefore

$$\begin{aligned} EE^T &= TT^T, \text{ since } XX^T = I \\ &= T_1^2 + \cdots + T_9^2 \end{aligned}$$

and

$$\frac{L_A^2}{6} + \cdots + \frac{(Q_A Q_B)^2}{36} = \sum T_i^2 - \frac{M^2}{9},$$

is the total variation. Thus the total variation can be split into variations due to $L_A, L_B, Q_A$, etc. each with 1 d.f.

**Yates' process. (r replications)**

| | | (i) | (ii) | Dividing Factors |
|---|---|---|---|---|
| $a_0b_0$ | $T_1$ | $S_1$ | $S_1' \to M$ | |
| $a_1b_0$ | $T_2$ | $S_2$ | $S_2' \to L_A$ | 6r |
| $a_2b_0$ | $T_3$ | $S_3$ | $S_3' \to Q_A$ | 18r |
| $a_0b_1$ | $T_4$ | $L_1$ | $L_1' \to L_B$ | 6r |
| $a_1b_1$ | $T_5$ | $L_2$ | $L_2' \to L_A L_B$ | 4r |
| $a_2b_1$ | $T_6$ | $L_3$ | $L_3' \to Q_A L_B$ | 12r |
| $a_0b_2$ | $T_7$ | $Q_1$ | $Q_1' \to Q_B$ | 18r |
| $a_1b_2$ | $T_8$ | $Q_2$ | $Q_2' \to L_A Q_B$ | 12r |
| $a_2b_2$ | $T_9$ | $Q_3$ | $Q_3' \to Q_A Q_B$ | 36r |

$$S_1 = T_1 + T_2 + T_3,\ \ S_2 = T_4 + T_5 + T_6,\ \ S_3 = T_7 + T_8 + T_9;$$

$$L_1 = T_3 - T_1,\ \ L_2 = T_6 - T_4,\ \ L_3 = T_9 - T_7 \text{ - (correspond to linear effects);}$$

$$Q_1 = T_3 - 2T_2 + T_1,\ \ Q_2 = T_6 - 2T_5 + T_4,\ \ Q_3 = T_9 - 2T_8 + T_7.$$

Repeat the process as many times as there are factors. Here there are only 2 factors and hence the process stops after 2 steps. The dividing factors are given by the sum of squares of the coefficients times the number of replications.

# Chapter 8

# Elements of Modern Algebra

In this chapter some relevant elements of modern algebra, used in the construction of designs, are given. The proofs of most of the theorems have been omitted.

**Definition 8.1**

In a finite group G the least positive integer r such that $a^r = e$, where e is the group identity, is defined as the order of the element.

**Theorem 8.1**

Let G be a finite group. For every element $a \in G$, an r exists such that $a^r = e$.

**Proof.**

Let g be the order of G and a any element. Consider the elements $a, a^2, \ldots, a^g, a^{g+1}$. They are all elements of the group and are $(g+1)$ in number. Hence there will be at least two elements in this set which are equal. Let

$$a^l = a^m, \ \ l > m.$$

Post – multiplying by $a^{-m}$, we have $a^{l-m} = e$, i.e., $a^r = e$, where $r = l - m$. ■

## 8.1 Cyclic Groups

A group whose elements can all be expressed as powers of a single element is called a cyclic group. If a is of order r, the cyclic group of order r generated by a is

$$(a) \equiv \{a, a^2, \ldots, a^{r-1}, a^r = e\}.$$

The order of a cyclic group is equal to the order of the generating element, and conversely, if a group of order r contains an element of order r, then the group is cyclic.

**Theorem 8.2**

A cyclic group is abelian.

**Theorem 8.3**

In any group, the order of $a \cdot b$ is equal to the order of $b \cdot a$.

**Theorem 8.4**

Order of the inverse is the same as the order of the element.

**Theorem 8.5**

If r is the order of a and if $a^m = e$, then $m \equiv 0$ (mod r).
If not, let $m = rq + s,\ \ 0 \leq s < r$. Then,

$$e = a^m = a^{rq+s} = (a^r)^q \cdot a^s = ea^s = a^s.$$

This contradicts the definition of r unless $s = 0$. Therefore $s = 0$.

**Theorem 8.6**

a and $a^t$ will be of the same order if and only if t is prime to r where r is the order of a.

**Case (i).**

Let r and t be prime to each other. Consider the elements

$$a^t, a^{2t}, \ldots, a^{(r-1)t}.$$

(Note that $a^{rt} = e$). The indices $t, 2t, \ldots, (r-1)t$ form an arithmetic progression and since t is prime to r, the remainders when they are divided by r are $1, 2, \ldots, r-1$, taken is some order. None of these numbers is a multiple of r and consequently, none of the elements $a^t, a^{2t}, \ldots, a^{(r-1)t}$ is equal to e. So r is the least positive integer having the property $a^{rt} = (at)^r = e$. So order of $a^t$ is r.

**Case (ii)**

Let t and r not be prime to each other and let c be the highest common factor of t and r. Let

$$t = ct', \quad r = cr'$$

where t' and r' are prime to each other.

Let $b = a^c$. Then $b^{r'} = a^{cr'} = a^r = e$. So r is the order of b. Again, $b^{t'} = a^{ct'} = a^t$. But t and r are prime to each other and r' is the order of b. Hence, by case (i), r is the order of $b^{t'} = a^t$.

Let G be a cyclic group of order g with a as generator. Then

$$G = \{a, a^2, \ldots, a^{g-1}, a^g = e\}.$$

All powers of a which are prime to g correspond to elements of order g. So the number of elements of order g is $\varphi(g)$, where $\varphi(n)$ is the Euler $\varphi$-function.

**Theorem 8.7**

The order of every element is a factor of the order of the group.

**Proof.**

Let $a_1$ be an element of order r of a group G whose order is g.

Consider the elements $a_1, a_1^2, \ldots, a_1^r = e$ which are all distinct. If this set exhausts G, then $g = r$ and the theorem is proved.

If this set does not exhaust the group, let $a_2$ be an element not included in the set. Consider the elements $a_2a_1, a_2a_1^2, \ldots, a_2a_1^r = a_2$. They are all in G. Further they are distinct also. For, if $a_2a_1^m = a_2a_1^n, m \neq n$, then $a_1^m = a_1^n$, which is not possible. Moreover, they are all different from the elements of the first set. For, If possible, let

$$a_1^l = a_2a_1^p.$$

If $l = p$, then $a_2 = e$ which is not true. If $l \neq p$, say $l > p$, then $a_1^{l-p} = a_2$, so that $a_2$ is an element of the first set, which is also not true.

If this second set, together with the first exhausts the group, then $g = 2r$ and the theorem is proved.

If not, take another element $a_3$ and we can show that r new elements also belong to G. Proceeding like this, since G is finite we will exhaust G at some stage, say the n-th. Then $g = nr$. ■

**Definition 8.2**

$n = \frac{g}{r}$ is called the index of $a_1$ in G.

**Complexes and Subgroups**

Let G be a group. Any subset of G is called a complex. A complex which is a group under the operation defined is a subgroup.

**Theorem 8.8**

The order of a subgroup is a factor of the order of the group.

**Proof.**

Let G be a finite group of order g, and H a subgroup. Let $H = \{a_1 = e, a_2, \dots, a_h\}$ where h is the order of H. If $H \neq G$, let $a'$ be an element of G not belonging to H. Consider the set $a_1a', \dots, a_ha'$. All these are in G, but not in H. They are distinct also. Thus we get 2h distinct elements of G. If $2h = g$, the theorem is proved.

If $2h \neq g$, let $a''$ be an element of G different from the above 2h elements. Consider the h elements $a_1a'', \dots, a_ha''$. As before, these are h distinct elements of G distinct from the above 2h elements. Thus we get a set of 3h distinct elements belonging to G. If this set exhausts the group then $3h = g$ and the theorem is proved.

If not, continue the process. Since G is finite the process will terminate after a finite number, say n, of steps and then $g = nh$.

$n = \frac{g}{h}$ is called the index of H in G. ■

**Theorem 8.9**

If a is of order r it will generate a subgroup of order r. (Consider the elements $a, a^2, \ldots, a^{g-1}, a^g = e$).

**Theorem 8.10**

A group of prime order is abelian.

**Proof.**

Let g be the order of G where g is a prime. The order of every element is a factor of g. Since g is prime the order of every element is either 1 or g. But e is the only element of order 1. So all other elements are of order g. Hence there is at least one element which will generate the group. Thus G is cyclic and hence abelian. ■

**Definition 8.3**

Let H be a subgroup of G. Then $Ha = \{ha: \ h \in H\}$ is called the right coset of a in G. Similarly $aH = \{ah: \ h \in G\}$ is called the left coset.

**Theorem 8.11**

If $a \in H$, then $Ha = aH = H$.

**Definition 8.4**

$a, b \in G$ are said to be conjugates if there exists an element $t \in G$ such that $b = t^{-1}at$.

**Theorem 8.12**

Conjugate elements are of the same order.

**Proof.**

Let a, b be conjugates. Then $b = t^{-1}at$ for some $t \in G$. Let r be the order of a and $r'$ the order of b. Then,

$$b^2 = t^{-1}at \cdot t^{-1}at = t^{-1}a^2t.$$

Similarly, $b^{r'} = t^{-1}at$ and $b^{r'} = e$, i.e., $t^{-1}a^{r'}t = e$. Therefore $a^{r'} = e$, and $r \leq r'$. Again, since b is conjugate to a, $a = tbt^{-1}$. Therefore

$$a^r = tbt^{-1} \text{ and } a^r = e.$$

$$tbt^{-1} = e \text{ or } b^r = e.$$

And $r' \leq r$. Therefore

$$r' = r. \blacksquare$$

**Theorem 8.13**

The elements of G which commute with a fixed element a of G is a subgroup.

**Definition 8.5**

This subgroup is called the normalizer of a in G, denoted by N(a).

**Example 8.1**

$$G = \left\{x, \frac{1}{x}, 1 - x, \frac{x}{x-1}, \frac{x-1}{x}\right\}.$$
$$N(1 - x) = \{x, 1 - x\}.$$

**Definition 8.6**

The complex which consists of all elements conjugate to a is called the class of a denoted by [a]. The class of a will contain the elements $tat^{-1}$ where t ranges over all G. Conjugacy is an equivalence relation. The class of a is then precisely the equivalence class of a under this equivalence relation.

**Theorem 8.14**

The number of elements in $[a]$ is h where $g = nh$ and n is the order of $N[a]$.

**Definition 8.7**

A subgroup H of G is called an invariant or normal or self-conjugate subgroup of G if $xH = Hx$ or $xHx^{-1} = H$ for every $x \in G$.

**Theorem 8.15**

Every subgroup of order n of a group of order 2n is an invariant subgroup.

**Alternating character**

Consider the determinant

$$D = \begin{vmatrix} 1 & 1 & \cdots & 1 \\ a_1 & a_2 & \cdots & a_n \\ a_1^2 & a_2^2 & \cdots & a_n^2 \\ \cdots & \cdots & \cdots & \cdots \\ a_1^{n-1} & a_2^{n-1} & \cdots & a_n^{n-1} \end{vmatrix},$$

and the permutation $P = (\lambda_i^i)$. If P operates on D, let it change to $D_P$. Then

$$D_P = \xi(P)D, \text{ where } \xi(P) = \pm 1.$$

$\xi(P)$ is called the alternating character of P.

A permutation P is said to be even or odd according as $\xi(P) = +1$ or $-1$. In any group of permutations G, either all or exactly half the permutations are even and the even permutations form a group.

**Homomorphism**

Let G and $G'$ be two groups. A mapping $\varphi: G \to G'$ is said to be a homomorphism if

$$\varphi(ab) = \varphi(a)\varphi(b) \quad \text{for all } a, b \in G.$$

[ ab on the left stands for multiplication in G and $\varphi(a)\varphi(b)$ on the right stands for multiplication in $G'$.]

**Theorem 8.16**

If $\varphi: G \to G'$ is a homomorphism, then $\varphi(e) = e'$ and $\varphi(a^{-1}) = \left(\varphi(a)\right)^{-1}$ for any $a \in G$.

**Isomorphism**

$\varphi: G \to G'$ is said to be an isomorphism if it is a one-to-one homomorphism of G onto $G'$.

**Theorem 8.17 (Cayley)**

Every finite group of order g is isomorphic to a permutation group with g objects.

**Theorem 8.18**

If H is an invariant subgroup of G, the set of right (left) cosets of H forms a group.

This group is called the quotient group of G relative to H and is denoted by $G/H$. If G is finite of order g and H is of order h, then order of $G/H$ is $\frac{g}{h} = n$, the index of H in G.

**Theorem 8.19**

If H is an invariant subgroup of G, then G is homomorphic to $G/H$.

**Proof.**

For $x \in G$, define $\varphi(x) = Hx$. Then $\varphi: G \to G/H$ is a homomorphism. ■

**Theorem 8.20**

Let $\varphi: G \to G'$ be a homomorphism. Then

$$K(\varphi) = \{a \in G: \ \varphi(a) = e'\}$$

is called the kernel of $\varphi$. $K(\varphi)$ is a normal subgroup of G.

**Rings**

Let R be a set of elements in which two operations called addition (+) and multiplication ($\cdot$) are defined. Let R be an abelian group w.r.t. addition. It is called a ring if in addition the following conditions hold:

(i) Closure w.r.t. multiplication

i.e., $a, b \in R \Rightarrow a \cdot b \in R$ for all $a, b \in R$.

(ii) Associativity w.r.t. multiplication.

For any 3 elements a, b, c of R,

$$(a \cdot b) \cdot c = a \cdot (b \cdot c).$$

(iii) Distributive Law.

If a, b, c are any 3 elements of R, then

$$a \cdot (b + c) = a \cdot b + a \cdot c$$

and

$$(b + c) \cdot a = b \cdot a + c \cdot a.$$

**Homomorphism**

Let R and R' be two rings. Then a mapping $\varphi: R \to R'$ is said to be a homomorphism if

$$\varphi(a \cdot b) = \varphi(a)\varphi(b)$$

and

$$\varphi(a + b) = \varphi(a) + \varphi(b)$$

for all $a, b \in R$.

If $\varphi$ is one-to-one, we say $\varphi$ is an isomorphism.

## 8.2 Ideals

$C \subset R$ is called an ideal of R if it is a subgroup of R under addition and for any element $c \in C$ and any element $r \in R$, rc and cr belong to C.

Every ideal M partitions a ring R into its distinct cosets. These cosets form a ring called the quotient ring of R relative to M.

### Integral Domain

A commutative ring R with the identity element for multiplication is called an integral domain, provided $a \cdot b = 0 \Rightarrow$ either $a = 0$ or $b = 0$, where 0 is the additive identity of R.

### Theorem 8.21

In an integral domain, the order of all non-zero elements under addition is the same.

The common order of the non-zero elements w.r.t. addition is called the characteristic of the integral domain. The characteristic of an integral domain is either a prime or is infinite.

### Field

An integral domain where the inverse w.r.t. multiplication of every non-zero element exists is called a field. A finite field is called a Galois field.

### Theorem 8.22

The number of elements in a Galois field is a prime or a power of a prime.

**Proof.**

Let F be a Galois field. Let p be the characteristic of F. Then p is a prime.

Let $a_1 \in F$. Then $r_1a_1, r_1 = 1,2, \dots, p$ are all distinct elements of the field $(pa_1 = 0)$. If this set exhausts the field, then the number of elements is p, a prime. If it does not exhaust the field, let $a_2$ be an element not included in this set. Consider the $p^2$ elements

$$r_1a_1 + r_2a_2; \quad r_1, r_2 = 1,2, \dots, p.$$

They are all distinct elements of F. If possible, let

$$r_ia_1 + r_ja_2 = r_la_1 + r_ma_2 .$$

Then $(r_i - r_l)a_1 = (r_m - r_j)a_2$. Here $r_i - r_l$ and $r_m - r_j$ are less than p and they are different also. Further $a_2$ is not a multiple of $a_1$. So

$$r_ia_1 + r_ja_2 \neq r_la_1 + r_ma_2 ,$$

unless $r_i = r_l$ and $r_m = r_j$.

If this set exhausts F we get $p^2$ elements in F. Otherwise, proceed as before. Since F is finite, at some stage all the elements will be exhausted. Let it happen at the n-th stage. Then F has $p^n$ elements. ■

## 8.3 Polynomial Rings

Let R be a ring. Let x be an "indeterminate" which commutes with every element of R. An expression of the form

$$P(x) = a_0 + a_1x + \dots + a_nx^n,$$

where the a's are elements of R and $a_n \neq 0$, is called a polynomial of degree n over R. Denote by $R[x]$ the set of all such polynomials. Under the familiar operations of addition and multiplication of polynomials, $R[x]$ is a ring, called the ring of polynomials over R.

### Irreducible Polynomials

Consider the polynomial ring $F[x]$ where F is a field. In fact $F[x]$ is an integral domain. Then $f(x) \in F[x]$ is said to be irreducible over F if it cannot be factored into factors of lower degree over F.

**Example 8.2**

Consider the filed $F = \{0,1\}$ under addition and multiplication modulo 2. Then the polynomial $x^2 + 1$ is reducible over F since

$$x^2 + 1 = x^2 + 2x + 1 = (x + 1)^2.$$

But over the field of real numbers, $x^2 + 1$ is irreducible. However $x^2 + x + 1$ is irreducible over F.

Next, consider $F = \{0,1,2\}$ with addition and multiplication modulo 3. Then

$$x^2 + x + 1 = x^2 + 4x + 4 = (x + 2)^2$$

so that $x^2 + x + 1$ is reducible over F.

**Theorem 8.23**

Let $\big(f(x)\big)$ denote the ideal of all multiples of $f(x) \in F[x]$ by elements of $F[x]$. Form the quotient ring $F[x]/f(x)$. Then $F[x]/f(x)$ is a field if and only if $f(x)$ is irreducible over F.

**Proof.**

Necessity.

Let $F[x]/f(x)$ be a field. If possible, let

$$f(x) = a_1(x)a_2(x)$$

where $a_1(x),\ a_2(x) \in F[x]$. Then

$$a_1(x) \cdot a_2(x) \equiv 0 \ \big(\text{mod}\, f(x)\big).$$

This is not possible since neither $a_1(x)$ nor $a_2(x)$ is equal to zero (mod $f(x)$). So $f(x)$ is irreducible over F.

Sufficiency.

Let $f(x)$ be irreducible over F. Already $F[x]$ is an integral domain. So $F[x]/f(x)$ is also an integral domain. Therefore, if we can prove that the inverse w.r.t. multiplication of all non-zero elements exist in $F[x]/f(x)$, then our assertion is proved.

Let $\varphi(x) \in F[x]/f(x)$. Then $f(x)$ and $\varphi(x)$ are relatively prime. (by definition.) So there exist two polynomials $l(x)$ and $h(x)$ such that

$$f(x) \cdot l(x) + \varphi(x) \cdot h(x) = 1 \pmod{f(x)}.$$

$$\text{i.e.,} \quad \varphi(x) \cdot h(x) = 1 - f(x) \cdot l(x) \pmod{f(x)}$$

$$= 1 \pmod{f(x)}.$$

Thus $h(x)$ is the inverse of $\varphi(x)$ and this completes the proof. ■

**Example 8.3**

(i) Let F be the real number field. Then $x^2 + 1$ is irreducible over F and $F[x]/f(x)$ will contain elements of the form $ax + b$ where $a, b$ are real numbers. $F[x]/f(x)$ is a field containing F as a subfield. (case when $a = 0$.)

$$(a_1x + b_1)(a_2x + b_2)$$
$$= a_1a_2x^2 + (a_1b_2 + a_2b_1)x + b_1b_2$$
$$= (a_1b_2 + a_2b_1)x + (b_1b_2 - a_1a_2) \pmod{x^2 + 1}.$$

Putting $a_1 = a_2 = 1$ and $b_1 = b_2 = 0$, we get $x^2 = -1$. Write $x = \sqrt{-1} = i$. Thus the new field we are getting is the set of all numbers of the form $ai + b$ where $a, b$ are real, i.e., the extended field is the complex number field.

(ii) Let $F = \{0,1\}$ under addition and multiplication modulo 2. In this field $x^2 + x + 1$ is irreducible and

$$\frac{F(x)}{(x^2 + x + 1)} = \{ax + b;\ a, b \in F\}$$
$$= \{0, 1, x, x + 1\}.$$

(iii) Let $F = \{0,1,2\}$ under addition and multiplication module 3. Then $x^2 + 1$ is irreducible over F and

$$\frac{F(x)}{(x^2 + 1)} = \{ax + b; a, b \in F\}.$$

If consists of $3^2$ elements.

In general, if $f(x)$ is an n-th degree polynomial which is irreducible in the mod p field (p prime) then the extended field given by $F[x]/f(x)$ will contain $p^n$ elements.

**Theorem 8.24**

For any prime $p\ (\geq 2)$ and every integer n (>0), there exists a Galois field with $p^n$ elements.

To prove this theorem, we need the following three lemmas.

**Lemma 8.1**

Every irreducible polynomial of degree n in the mod p field is a divisor of $x^{p^n-1} - 1$.

**Proof.**

Let $f(x)$ be an irreducible polynomial of degree n in the mod p field. Let

$$f(x) = a_0 + a_1x + \cdots + a_nx^n.$$

Then $F[x]/f(x)$ is a Galois field of order $p^n$ and the number of non-zero elements in this field is $p^n - 1$. But we know that the non-zero elements in a field form a multiplicative group of order $p^n - 1$. Let x be any element of this group. Then

$$x^{p^n-1} \equiv 1 \mod p\,,$$

and

$$x^{p^n-1} - 1 = 0 \mod p.$$

Therefore $f(x)$ is a factor of $x^{p^n-1} - 1$. ∎

**Lemma 8.2**

No irreducible polynomial of degree $r > n$ is a factor of $x^{p^n-1} - 1$.

**Proof.**

Let f(x) be an irreducible polynomial of degree r (>r) in the mod p field. If possible, let f(x) be a divisor of $x^{p^n-1}-1$. Consider the Galois field $F[x]/f(x)$ consisting of $p^r$ elements, the number of non-zero elements being $p^r-1$. Every element in this field is of the form

$$a_0+a_1x+\cdots+a_kx^k, k<r,$$

where $a_0, a_1, \ldots, a_k$ are residues mod p. Then

$$\begin{aligned}(a_0+a_1x+\cdots+a_kx^k)^{p^n}& \\ &=a_0^{p^n}+(a_1x)^{p^n}+\cdots+(a_kx^k)^{p^n} \pmod p\\ &=a_0+a_1x+\cdots+a_kx^k \pmod p,\end{aligned}$$

since $x^{p^n}=x \pmod p$ by assumption. [In the mod p field, $(a\pm b)^p=a^p\pm b^p$.]
Therefore

$$(a_0+a_1x+\cdots+a_kx^k)^{p^n-1}=1 \mod (f(x),p).$$

This shows that the order of every element in $F[x]/f(x)$ is $p^n-1$. Therefore there cannot be more than $p^n-1$ non-zero elements in this field. This contradiction proves the lemma. ■

**Lemma 8.3**

$x^{p^n-1}-1$ has no repeated roots.

**Proof.**

Differentiating $x^{p^n-1}-1=0$ w.r.t. x, we get

$$(p^n-1)x^{p^n-2}=0,$$

and since these two equations cannot have a common root, the factors of $x^{p^n-1}-1$ are all distinct. ■

**Proof of Theorem 8.24**

The polynomial $x^{p^n-1} - 1$ has no irreducible factor of degree larger than n in the mod p field. All the irreducible polynomials of degree m (<n) are factors of $x^{p^m-1} - 1$. Since $x^{p^n-1} - 1$ has no double root, the sum of the degrees of all irreducible factors of degree m together is less than or equal to $p^m - 1$. Hence the sum of the degrees of all factors of degree less than n is at most

$$\sum_{1}^{n-1} p^m = \frac{p^n - 1}{p - 1},$$

and the degree of the function is $p^n - 1$. Therefore there must be at least one irreducible factor of degree n. Let $f(x)$ be this factor. Then $F[x]/f(x)$ is a Galois field of order $p^n$. ■

Denote this Galois field by $GF(p^n)$.

**Primitive Element**

Let there be $s = p^n$ elements in a field. If $\alpha$ is an element of the field such that $\alpha, \alpha^2, \ldots, \alpha^{s-1}$ are the non-zero elements of the field, then $\alpha$ is called a primitive element of the field. If $\alpha$ is a root of

$$x^{p^n-1} - 1 \equiv 0 \bmod (f(x), p),$$

then $\alpha$ is a primitive element and all the non-zero elements of the field can be expressed as powers of $\alpha$, i.e. $\alpha, \alpha^2, \ldots, \alpha^{p^n-1}$.

**Example 8.4**

Consider the elements of the Galois field mod $(x^3 + x + 1, 2)$.

So $x^3 + x + 1$ is irreducible in the mod 2 field and the residue class will form a field with $2^3 = 8$ elements. Let the remainders be of the form

$$ax^2 + bx + c$$

where $a, b, c$ are residues mod 2. The 8 elements of the extended field are given by

$$0, 1, x, x+1, x^2+x+1, x^2+1, x^2, x^2+x.$$

Now,

$$\begin{aligned} x &= x \\ x^2 &= x^2 \\ x^3 &= x+1 \\ x^4 &= x^2+x \\ x^5 &= x^3+x^2 = x^2+x+1 \\ x^6 &= x^3+x^2+x = x^2+1 \\ x^7 &= x^3+x = 2x+1 = 1. \end{aligned}$$

Therefore x is a primitive element of this field.

**Definition 8.8**

An element of order $p^n - 1$ in the Galois field of order $p^n$ is called a primitive root.

**Theorem 8.25**

The Galois field $GF(p^n)$ has $\varphi(p^n - 1)$ primitive roots, where $\varphi(r)$ denotes the number of numbers less than r and prime to r.

**Proof.**

Let s be the highest order of the elements in $GF(p^n)$. Then, since the order of every element divides s, every element in $GF(p^n)$ satisfies the equation

$$x^s = 1 \quad \text{or} \quad x^s - 1 = 0. \tag{1}$$

Consider the multiplicative group of order $p^n - 1$ formed by the $p^n - 1$ non-zero elements. For every $\alpha$ in this group, we have

$$\alpha^{p^n-1} - 1 = 0.$$

i.e., every $\alpha$ in the group is a root of

$$x^{p^n-1} - 1 = 0. \tag{2}$$

Now, the number of distinct roots of equation (1) is at most s.

But by (2), $p^n - 1$ distinct elements satisfy this equation. Therefore,

$$s \geq p^n - 1. \tag{3}$$

Again, the order of any element cannot exceed $p^n - 1$. This gives

$$s \leq p^n - 1. \tag{4}$$

From (3) and (4), $s = p^n - 1$. So there is at least one element of order $p^n - 1$, i.e., there is at least one primitive element. Let it be w. Then $w^t$, where t is prime to $p^n - 1$ is also a primitive element. Thus there are $\varphi(p^n - 1)$ primitive elements. ■

**Theorem 8.26**

Any primitive element of $GF(p^n)$ will satisfy an irreducible polynomial of degree n.

**Proof.**

Let $\alpha$ be a primitive element of $GF(p^n)$. Then the elements of the field will be $\alpha^0 = 1, \alpha^1, \alpha^2, \ldots, \alpha^{s-1}$ where $s = p^n$. Any element of the field can be expressed in the form

$$\sum_{j=1}^{n} r_j v_j$$

where $r_1, \ldots, r_n$ are all residues mod p and the v's are elements of the field. Now $\alpha$ is a primitive element. So

$$\alpha^i = \sum_{j=1}^{n} r_{ij} v_j, \text{ say.}$$

Consider the $(n+1)$ elements.

$$\alpha^0 = r_{01}v_1 + r_{02}v_2 + \cdots + r_{0n}v_n,$$

$$\alpha^1 = r_{11}v_1 + r_{12}v_2 + \cdots + r_{1n}v_n,$$

$$\cdots\cdots\cdots\cdots\cdots\cdots\cdots\cdots\cdots\cdots\cdots\cdots\cdots\cdots$$

$$\alpha^n = r_{n1}v_1 + r_{n2}v_2 + \cdots + r_{nn}v_n.$$

Eliminating $v_1, v_2, \ldots, v_n$, we will get an equation of the form

$$a_0 + a_1\alpha + a_2\alpha^2 + \cdots + a_n\alpha^n = 0.$$

Thus $\alpha$ satisfies the n-th degree polynomial

$$a_0 + a_1x + a_2x^2 + \cdots + a_nx^n.$$

We have to show that it is irreducible. If it is reducible, let

$$b_0 + b_1x + \cdots + b_kx^k, \quad (k < n, b_k \neq 0)$$

be the lowest degree factor. Then

$$b_0 + b_1\alpha + \cdots + b_k\alpha^k = 0.$$

Therefore

$$b_k\alpha^k = -b_0 - b_1\alpha - \cdots - b_{k-1}\alpha^{k-1}.$$

Premultiplying by $b_k^{-1}$, $\alpha^k$ can be expressed in terms of lower powers of $\alpha$. So all other elements can be expressed as linear combinations of $\alpha^0, \alpha^1, \alpha^2, \ldots, \alpha^{k-1}$ where the compounding coefficients will be residues mod p. Hence the maximum number of non-zero elements in $GF(p^n)$ is $p^{k-1}$. This is a contradiction. ■

**Theorem 8.27 (Converse of Theorem 8.26)**

If $\alpha$ satisfies an irreducible polynomial of degree n, then it is a primitive root.

**Proof.**

Let $a_0 + a_1x + a_2x^2 + \cdots + a_nx^n$ be an irreducible polynomial and let

$$a_0 + a_1\alpha + a_2\alpha^2 + \cdots + a_n\alpha^n = 0.$$

Then

$$\alpha^n = c_0 + c_1\alpha + c_2\alpha^2 + \cdots + c_{n-1}\alpha^{n-1},$$

where

$$c_0 = -a_n^{-1}a_0 \pmod{p},$$
$$c_1 = -a_n^{-1}a_1 \pmod{p}, \text{ etc..}$$

By giving possible values to $c_0, c_1, \ldots, c_{n-1}$, viz., $0,1,\ldots,p-1$, we get all the $p^n$ distinct elements of the field. So $\alpha$ is a primitive element. ■

## 8.4 Cyclotomic Polynomial

If x in the field $F[x]/f(x)$, where $f(x)$ is an irreducible polynomial of degree n over the field $F = \{0,1,\ldots,p-1 \pmod{p}\}$, does not satisfy any equation $x^m - 1 = 0$ with $m < p^n - 1$, then x is a primitive root. On the other hand, if $\alpha$ is a primitive root of $GF(p^n)$ then $\alpha$ must satisfy an irreducible equation of degree n. Thus, if we wish to have $GF(p^n)$ presented by the residues mod $(f(x), p)$ in such a way that x is a primitive root, then we have to remove from $x^{p^n-1} - 1$ all factors which are factors of $x^m - 1$ for any $m < p^n - 1$. The remaining polynomial has as its roots all the primitive roots of $GF(p^n)$ and since the total number of primitive roots is $\varphi(p^n - 1)$, its degree will be $\varphi(p^n - 1)$.

**Definition 8.9**

A polynomial with all its roots primitive (of order $p^n - 1$) is called a cyclotomic polynomial of order $p^n - 1$.

**Construction**

(i) Consider $GF(2^3)$. We form the cyclotomic polynomial of order $2^3 - 1 = 7$, whose degree is $\varphi(2^3 - 1) = \varphi(7) = 6$. Removing the root 1 from $x^7 - 1$ we get

$$x^6 + x^5 + x^4 + x^3 + x^2 + x + 1.$$

This polynomial must decompose into 2 factors of degree 3 each in the mod 2 system. Thus if

$$(x^6 + x^5 + x^4 + x^3 + x^2 + x + 1) \equiv (x^3 + a_1x^2 + b_1x + c_1)(x^3 + a_2x^2 + b_2x + c_2),$$

then $c_1c_2 = 1$ mod 2, so that $c_1 = c_2 = 1$, and $b_1 + b_1 = 1$ mod 2. Let $b_1 = 0$ and $b_2 = 1$. Then

$$a_2c_1 + a_1c_2 + b_1b_2 = a_1 + a_2 = 1 \pmod 2,$$

$$c_1 + c_2 + a_1b_2 + a_2b_1 = a_1 = 1 \pmod 2.$$

Therefore $a_1 = 1$ and $a_2 = 0$. Thus

$$(x^6 + x^5 + x^4 + x^3 + x^2 + x + 1) \equiv (x^3 + x^2 + 1)(x^3 + x + 1) \pmod 2.$$

Therefore there are two irreducible polynomials of degree 3 in the mod 2 field and they are

$$f_1(x) = x^3 + x^2 + 1 \quad \text{and} \quad f_2(x) = x^3 + x + 1.$$

Hence the non-zero elements of $GF(2^3)$ in terms of x are given by:

| $f_1(x) = x^3 + x^2 + 1$ | $f_2(x) = x^3 + x + 1$ |
|---|---|
| $x = x,$ | $x = x,$ |
| $x^2 = x^2,$ | $x^2 = x^2,$ |
| $x^3 = x^2 + 1,$ | $x^3 = x + 1,$ |
| $x^4 = x^3 + x = x^2 + x + 1,$ | $x^4 = x^2 + x,$ |
| $x^5 = x^3 + x^2 + x = x + 1,$ | $x^5 = x^2 + x + 1,$ |
| $x^6 = x^2 + x,$ | $x^6 = x^2 + 1,$ |
| $x^7 = x^3 + x^2 = 1.$ | $x^7 = 1.$ |

(ii) Consider $GF(5^2)$. The degree of the cyclotomic polynomial of order $5^2 - 1 = 24$ is $\varphi(5^2 - 1) = \varphi(24) = 8$. We have

$$x^{24} - 1 = (x^{12} - 1)(x^{12} + 1).$$

The factor $x^{12} + 1$ has to be split into 2 factors, one of degree 4 and the other of degree 8 and the quadratic factors of the polynomial of degree 8 give the irreducible polynomials of degree 2. We have

$$x^{12} + 1 = (x^4 + 1)(x^8 - x^4 + 1).$$

So we have to split $x^8 - x^4 + 1$ into factors of degree 2 each. We have

$$x^8 - x^4 + 1 = (x^2 + 2x + 3)(x^6 - 2x^5 + x^4 - x^3 - 2x^2 + 2x - 3),$$

and so one irreducible polynomial of degree 2 is $x^2 + 2x + 3$.

**Theorem 8.28**

Any tow Galois fields with the same number of elements are isomorphic. [In an abstract sense we therefore have only one Galois field with $p^n$ elements. It is this unique Galois field that is denoted by $GF(p^n)$.]

**Proof.**

It is easy to see that every $GF(p)$ is isomorphic with the system of residues mod p. We know that there exists a mod p irreducible polynomial $g(x)$ of degree n. Let F be any Galois field with $p^n$ elements which are $\alpha_0 = 0, \alpha_1 = 1, \alpha_2, \ldots, \alpha_{p^n-1}$. Then

$$x^{p^n-1} - 1 = (x - \alpha_1)(x - \alpha_2) \ldots (x - \alpha_{p^n-1}).$$

Since $g(x)$ is a divisor of $x^{p^n-1} - 1$ it follows that for some i we must have $g(\alpha_i) = 0$. Now g is irreducible, therefore the expressions

$$a_0 + a_1\alpha_i + a_2\alpha_i^2 + \cdots + a_{n-1}\alpha_i^{n-1}, \tag{5}$$

where the $a_i$ are multiples of the unit element of F must all be different from zero and thus also different from each other. Otherwise $g(x)$ would have a factor in common with a polynomial of degree less than n. Thus (5) presents $p^n$ different elements of F and hence all the elements of F. But the correspondence $f(\alpha_i) \leftrightarrow f(x)$ where $f(\alpha_i) \in F$ and $f(x) \in \bar{F} = F[x]/g(x)$ is clearly an isomorphism. Thus any tow fields F, F′ both having $p^n$ elements are isomorphic to $\bar{F}$ and hence isomorphic to each other. ∎

# Chapter 9

# Construction of Designs

## 9.1 Orthogonal Latin Squares

A Latin square of side m is an arrangement of m letters into $m^2$ subsquares of a square in such a way that every row and every column contains every letter exactly once. Two Latin squares are termed orthogonal if when one is super-imposed on the other, every ordered pair of symbols occurs exactly once in the resulting square. Thus the latin squares

$$\begin{matrix} A & B & C \\ B & C & A \\ C & A & B \end{matrix} \qquad \begin{matrix} \alpha & \beta & \gamma \\ \gamma & \alpha & \beta \\ \beta & \gamma & \alpha \end{matrix}$$

are orthogonal. The problem of constructing, for instance, a set of r orthogonal Latin squares of side m could be regarded as solved if we either can give a method by which such a design can be constructed or are able to prove that the design cannot exist. This problem is at present still unsolved for many combinations of r and m. Various methods have been discovered however for obtaining solutions in a great many cases. In fact, within the range useful in the design of experiments, the solution has been obtained for most cases with only a few exceptions. The experimenter will not usually go beyond $m = 13$. The problem of the construction of orthogonal latin squares within this range is solved for $m = 2,3,4,5,6,7,8,9,11,13$. That is to say, $(m - 1)$ orthogonal Latin squares of side m can be constructed for $m = 2,3,4,5,6,7,8,9,11,13$ while it has been proved that no six-sided orthogonal pair exists nor more than $(m - 1)$ orthogonal Latin squares of side m.

**Theorem 9.1**

The maximum number of orthogonal Latin squares of side s is $(s-1)$.

Consider a Latin square of side s. There will be a number of Latin squares which are all orthogonal to this Latin square. Let us superimpose all these Latin squares on to this Latin square. Let the first row of this superimposed Latin square be

| 111 ... | 222 ... | ... ... | sss ... |
|---|---|---|---|

Since all the Latin squares are orthogonal one letter combination appears once and only once in any cell. The number of distinct elements possible in a cell is s and out of these s one already appears in a cell of the first row. Therefore there remain only $(s-1)$ distinct letter combinations and this is the total number of orthogonal Latin squares possible. Therefore if all the Latin squares of side s exist, then maximum number is $(s-1)$.

**Theorem 9.2**

If s is a prime or the power of a prime all the orthogonal Latin squares can be constructed.

Let $s = p^n$ where p is a prime. Consider $GF(p^n)$. Let the elements be

$$\alpha_0 = 0, \alpha_1 = 1, \alpha_2, \alpha_3, \dots, \alpha_{s-1},$$

and let x be a primitive element of this field. Then the elements can be written in the form

$$\alpha_0 = 0, \alpha_1 = 1, \alpha_2 = x, \alpha_3 = x^2, \dots, \alpha_{s-1} = x^{s-2}.$$

Let us consider an $s \times s$ square arrangement $L_i$ of letters, where the $(p, q)$-th cell is filled by the letter j where j is defined by

$$\alpha_j = \alpha_{p-1} + \alpha_i \alpha_{q-1}.$$

For $i = 1,2, \dots, s-1$, we get $(s-1)$ such distinct squares.

**To Prove that $L_i$ is a Latin Square**

We need to prove that the letter in the $(p, q)$-th cell is different from the letter in the $(p', q')$-th cell. Taking a particular row, if possible let the letters in the $(p, q)$-th and $(p, q')$-th cells be the same, say j. Then

$$\begin{aligned}\alpha_j &= \alpha_{p-1} + \alpha_i \alpha_{q-1} \\ &= \alpha_{p-1} + \alpha_i \alpha_{q'-1}.\end{aligned}$$

Therefore $\alpha_i(\alpha_{q-1} - \alpha_{q'-1}) = 0$ giving $\alpha_{q-1} = \alpha_{q'-1}$, since $\alpha_i \neq 0$. This is possible only if $q = q'$.

Again, considering the $q$-th column, if possible let the $(p, q)$-th and the $(p', q)$-th letters be the same, say j. Then, as before, $\alpha_{p-1} = \alpha_{p'-1}$ or $p = p'$.

Combining the two results, the letters in the $(p, q)$-th and $(p', q')$-th cells are different.

Therefore $L_i$ is a Latin square.

Let us consider two Latin squares $L_i$ and $L_j$, $i \neq j$, and superimpose $L_i$ on $L_j$. Let $(l, m)$ be the observation in the $(p, q)$-th cell, where l is the letter in the $(p, q)$-th cell of $L_i$ and m is the letter in the $(p, q)$-th cell of $L_j$. Then

$$\begin{aligned}\alpha_l &= \alpha_{p-1} + \alpha_i \alpha_{q-1} \\ \alpha_m &= \alpha_{p-1} + \alpha_j \alpha_{q-1}.\end{aligned}$$

We shall prove that $L_i$ and $L_j$ are orthogonal. If possible, let the $(p, q)$-th and $(p', q')$-th cells of the superimposed square be the same, viz. $(l, m)$. Then, in $L_i$, the $(p, q)$-th and $(p', q')$-th cells are the same, viz. l, and similarly in $L_j$ also the $(p, q)$-th and $(p', q')$-th cells are the same, viz. m. Then only in the superimposed square the $(p, q)$-th and $(p', q')$-th cells be the same, viz. $(l, m)$. Thus

$$\alpha_l = \alpha_{p-1} + \alpha_i \alpha_{q-1} = \alpha_{p'-1} + \alpha_i \alpha_{q'-1},$$

and

$$\alpha_m = \alpha_{p-1} + \alpha_j \alpha_{q-1} = \alpha_{p'-1} + \alpha_j \alpha_{q'-1},$$

giving

$$\alpha_{p-1} - \alpha_{p'-1} = \alpha_i(\alpha_{q'-1} - \alpha_{q-1}),$$

and

$$\alpha_{p-1} - \alpha_{p'-1} = \alpha_j(\alpha_{q'-1} - \alpha_{q-1}).$$

Therefore

$$(\alpha_i - \alpha_j)(\alpha_{q'-1} - \alpha_{q-1}) = 0.$$

But $\alpha_i \neq \alpha_j$. So $\alpha_{q'-1} = \alpha_{q-1}$ or $q = q'$. Similarly, $p = p'$.

Thus different cells have different observations. Hence $L_i$ and $L_j$ are orthogonal.

Hence for $i = 1,2,\ldots,s-1$, all the Latin squares are mutually orthogonal. Therefore we can construct $(s-1)$ mutually orthogonal Latin squares.

**Key Latin Square and Key Line**

Consider the Latin square $L_1$. Let the $(p, q)$-th cell observation be j. Then

$$\alpha_j = \alpha_{p-1} + \alpha_1\alpha_{q-1} = \alpha_{p-1} + \alpha_{q-1}, \text{ since } \alpha_1 = 1.$$

Therefore the $(p, q)$-th cell observation is the same as the $(q, p)$-th cell observation. Therefore $L_1$ is symmetric.

$L_1$ is called the key Latin square and its second row is called the key line.

**Theorem 9.3**

The $q$-th column of $L_{i+1}$ is the same as the $(q+1)$-th column of $L_i$ and the last column of $L_{i+1}$ is the same as the second column of $L_i$.

Let us consider the $p$-th rows of $L_i$ and $L_{i+1}$. The observation in the $q$-th column of $L_{i+1}$ is given by

$$\begin{aligned}\alpha_{p-1} + \alpha_{i+1}\alpha_{q-1} &= \alpha_{p-1} + \alpha_i x \alpha_{q-1} \\ &= \alpha_{p-1} + \alpha_i\alpha_q,\end{aligned}$$

which gives the observation in the $(q+1)$-th column of $L_i$.

This is true only for $q \geq 2$. When $q = 1$, the second column of $L_i$ is given by

$$\alpha_{p-1} + \alpha_i \alpha_1 = \alpha_{p-1} + \alpha_i$$

and the last column of $L_{i+1}$ is given by

$$\begin{aligned} \alpha_{p-1} + \alpha_{i+1}\alpha_{s-1} &= \alpha_{p-1} + \alpha_{i+1}x^{s-2} \\ &= \alpha_{p-1} + \alpha_i x x^{s-2} \\ &= \alpha_{p-1} + \alpha_i x^{s-1} \\ &= \alpha_{p-1} + \alpha_i, \ \text{ since } \ x^{s-1} = 1. \end{aligned}$$

Therefore the last column of $L_{i+1}$ = second column of $L_i$.

Thus by constructing $L_1$, all other Latin squares can be constructed in a cyclic manner. But $L_1$ is symmetric and so by constructing half the letters of $L_1$ all other Latin squares can be constructed.

The key line is given by (here $p = 2$),

$$\begin{aligned} \alpha_{2-1} + \alpha_1 \alpha_{q-1} &= \alpha_1 + \alpha_{q-1} \\ &= 1 + \alpha_{q-1}. \end{aligned}$$

For $q = 1,2,\dots,s-1$, we get all the letters in the key line.

**Theorem 9.4**

The $(p+1, q+1)$-th cell observation in $L_1$ is the same as $1 + (p,q)$-th cell observation in $L_1$ if the $(p,q)$-th cell observation is different from zero.

Le the $(p,q)$-th cell boservaton be j. Then

$$\alpha_j = \alpha_{p-1} + \alpha_{q-1}.$$

Now, $\alpha_p + \alpha_q = x(\alpha_{p-1} + \alpha_{q-1}) = x\alpha_j = \alpha_{j+1}$. So, $x\alpha_j = 0$ if $\alpha_j = 0 = \alpha_0$. i.e., if the $(p,q)$-th cell is zero then the $(p+1, q+1)$-th cell is also zero.

Let $\alpha_j$ be the last element, viz. $x^{s-2}$.

$$\text{i.e.,} \quad \alpha_j = x^{s-2} = \alpha_{s-1}.$$

Therefore

$$x\alpha_j = x^{s-1} = 1 = \alpha_1.$$

Hence, if the $(p, q)$-th cell observation is $(s-1)$, then the $(p+1, q+1)$-th cell observation is 1.

In all other cases, the $(p+1, q+1)$-th cell observation is $1 + j = 1 + (p, q)$-th cell observation.

Thus any row of $L_1$ can be constructed from the previous row. The first row we write in the natural order, and the second row is given by the key line. Therefore, using the symmetry of $L_1$, the remaining rows also can be constructed.

**To Construct the Orthogonal Latin Quares of Side 4**

We have to construct $GF(2^2)$. We first form the cyclotomic polynomial of order $2^2 - 1 = 3$. Its degree is $\varphi(2^2 - 1) = \varphi(3) = 2$. Removing the root 1 from $x^3 - 1$ we get $x^2 + x + 1$ and this is the only irreducible polynomial of degree 2 in the mod 2 field. Hence the elements of $GF(2^2)$ are

$$\alpha_0 = 0, \alpha_1 = 1, \alpha_2 = x, \alpha_3 = x^2 = x + 1.$$

The key line of $L_1$ is $1 + \alpha_{q-1}$:

| Values of q | $1 + \alpha_{q-1}$ | Observation |
|---|---|---|
| 1 | $1 + \alpha_0 = 1 = \alpha_1$ | 1 |
| 2 | $1 + \alpha_1 = 2 = 0$ $= \alpha_0$ | 0 |
| 3 | $1 + \alpha_2 = 1 + x$ $= \alpha_3$ | 3 |
| 4 | $1 + \alpha_3 = x + 2$ $= x = \alpha_2$ | 2 |

So the key line is

| 1 | 0 | 3 | 2 |
|---|---|---|---|

and $L_1$ is

| 0 | 1 | 2 | 3 |
|---|---|---|---|
| 1 | 0 | 3 | 2 |
| 2 | 3 | 0 | 1 |
| 3 | 2 | 1 | 0 |

The other two Latin squares are

$L_2$:

| 0 | 2 | 3 | 1 |
|---|---|---|---|
| 1 | 3 | 2 | 0 |
| 2 | 0 | 1 | 3 |
| 3 | 1 | 0 | 2 |

$L_3$:

| 0 | 3 | 1 | 2 |
|---|---|---|---|
| 1 | 2 | 0 | 3 |
| 2 | 1 | 3 | 0 |
| 3 | 0 | 2 | 1 |

**To Construct all 8x8 Orthogonal Latin Squares**

We consider $GF(2^3)$. The cyclotomic polynomial of order $2^3 - 1 = 7$ and degree $\varphi(7) = 6$ is

$$x^6 + x^5 + x^4 + x^3 + x^2 + x + 1.$$

The two cubic factors of this polynomial are the two irreducible polynomials of degree 3 in the mod 2 field. Let

$$x^6 + x^5 + x^4 + x^3 + x^2 + x + 1 = (x^3 + a_1x^2 + b_1x + c_1)(x^3 + a_2x^2 + b_2x + c_2)$$

Then $c_1c_2 = 1$ (mod 2) so that $c_1 = c_2 = 1$ (mod 2), and

$$b_1c_2 + b_2c_1 = 1 \pmod 2.$$

$$\text{i.e.,} \quad b_1 + b_2 = 1 \pmod 2.$$

Let $b_1 = 0 \pmod 2$. Then $b_2 = 1 \pmod 2$ and

$$a_2c_1 + a_1c_2 + b_1b_2 = a_1 + a_2 = 1 \pmod 2,$$
$$c_1 + c_2 + a_1b_2 + a_2b_1 = a = 1 \pmod 2.$$

Therefore $a_1 = 1 \pmod 2$, and $a_2 = 0 \pmod 2$. And

$$x^6 + x^5 + x^4 + x^3 + x^2 + x + 1 = (x^3 + x^2 + 1)(x^3 + x + 1)$$

in the mod 2 field. Therefore $x^3 + x^2 + 1$ and $x^3 + x + 1$ are the two irreducible polynomials of degree 3 in the mod 2 field and any of them will give all the 7 non-zero elements of $GF(2^3)$. Taking $x^3 + x^2 + 1$, the elements are

$$\alpha_0 = 0, \alpha_1 = 1, \alpha_2 = x, \alpha_3 = x^2, \alpha_4 = x^3 = x + 1, \alpha_5 = x^2 + x,$$
$$\alpha_6 = x^3 + x^2 = x^2 + x + 1, \alpha_7 = x^3 + x^2 + x = x^2 + 1.$$

The key line is obtained as follows:

| Values of q | $1 + \alpha_{q-1}$ | Observation |
|---|---|---|
| 1 | $1 + \alpha_0 = 1 = \alpha_1$ | 1 |
| 2 | $1 + \alpha_1 = 2 = 0 = \alpha_0$ | 0 |
| 3 | $1 + \alpha_2 = 1 + x = \alpha_4$ | 4 |
| 4 | $1 + \alpha_3 = x^2 + 1 = \alpha_7$ | 7 |
| 5 | $1 + \alpha_4 = x + 2 = x = \alpha_2$ | 2 |
| 6 | $1 + \alpha_5 = x^2 + x + 1 = \alpha_6$ | 6 |
| 7 | $1 + \alpha_6 = x^2 + x = \alpha_5$ | 5 |
| 8 | $1 + \alpha_7 = x^2 = \alpha_3$ | 3 |

Hence

| 1 | 0 | 4 | 7 | 2 | 6 | 5 | 3 |
|---|---|---|---|---|---|---|---|

is the key line. The key Latin square $L_1$ is

| 0 | 1 | 2 | 3 | 4 | 5 | 6 | 7 |
|---|---|---|---|---|---|---|---|
| 1 | 0 | 4 | 7 | 2 | 6 | 5 | 3 |
| 2 | 4 | 0 | 5 | 1 | 3 | 7 | 6 |
| 3 | 7 | 5 | 0 | 6 | 2 | 4 | 1 |
| 4 | 2 | 1 | 6 | 0 | 7 | 3 | 5 |
| 5 | 6 | 3 | 2 | 7 | 0 | 1 | 4 |
| 6 | 5 | 7 | 4 | 3 | 1 | 0 | 2 |
| 7 | 3 | 6 | 1 | 5 | 4 | 2 | 0 |

$L_2, L_3, \ldots, L_7$ can be constructed from this.

**Construction of A Latin Square of Side s where s is a Prime**

Consider the field

$$F: 0,1,2, \ldots, s-1 \pmod{s}.$$

Fill up the $(p, q)$-th cell in $L_i$ by j where

$$\alpha_j = \alpha_{p-1} + \alpha_i \alpha_{q-1},$$

and the $\alpha$'s being such that

$$\alpha_0 = 0, \alpha_1 = 1, \alpha_2 = 2, \ldots, \alpha_{s-1} = s-1.$$

Thus $j = (p-1) + (q-1)i$. Consider the $p$-th row. When

$$q = 1, \;\; j = p - 1,$$

$$q = 2, \;\; j = (p-1) + i,$$

$$q = 3, \;\; j = (p-1) + 2i,$$

$$\ldots \ldots \ldots \ldots \ldots \ldots \ldots \ldots \ldots$$

$$q = s-1, \;\; j = (p-1) + (s-2)i,$$

$$q = s,\ \ j = (p-1) + (s-1)i\,,$$

i.e., each element is obtained by adding i to the previous element. So each column is obtained by adding i to the previous column, the addition being residue mod p.

By letting $p = 1,2,\dots,s$, we get $L_i$:

| 0 | i | 2i | ... | (s − 1)i |
|---|---|---|---|---|
| 1 | 1 + i | 1 + 2i | ... | 1 + (s − 1)i |
| 2 | 2 + i | 2 + 2i | ... | 2 + (s − 1)i |
| ... | ... | ... | ... | ... |
| s − 1 | (s − 1) + i | (s − 1) + 2i | ... | (s − 1) + (s − 1)i |

$L_i$ and $L_j$ are orthogonal for all i and j if $i \neq j$.

## 9.2 Construction of BIBD's

Consider a BIBD with parameters $b, k, v, r, \lambda$. Then for the complementary BIBD,

$$b' = b, k' = v - k, v' = v, r' = b - r, k + k' = v.$$

Hence given v we need construct only the complementary design for which $k' \leq v/2$, i.e., we need to construct a design with $k \leq v/2$.

$k = 2$ is only a trivial case, for we can construct the design for all values of v in this case.

$$b = \binom{v}{2} = \frac{v(v-1)}{2}.$$

$bk = rv$ gives $r = v - 1$. Similarly $r(k-1) = \lambda(v-1)$ gives $\lambda = 1$. Therefore we need consider cases where $3 \leq k \leq v/2$.

### Block Section and Intersection of Symmetrical BIBD's

### Yates Series A and Series B

This method is obtained by establishing a correspondence between BIBD's and orthogonal Latin squares.

Consider a complete system of Latin squares. If s is a prime or a power of a prime, there are $(s-1)$ mutually orthogonal Latin squares of side s. The correspondence is established as follow:

Cell in a Latin square → Treatment in a BIBD.

Column in a Latin square → Block in a BIBD.

Row in a Latin square → Block in a BIBD.

Letter in a Latin square → Block in a BIBD.

By this correspondence, we have

A cell in a column → Treatment in a block

A cell in a row → Treatment in a block

A cell containing a letter → Treatment in a block

Consider the $(s-1)$ mutually orthogonal Latin squares of side s, and consider the arrangement with all these $(s-1)$ designs superimposed. By the correspondence we have established,

$$\text{Total number of treatments} \\ = \text{Total number of cells in a Latin square} = s^2.$$

$$\text{i.e.,} \quad v = s^2.$$

$$\begin{aligned} b &= \text{Total number of blocks in a BIBD} \\ &= (\text{Total number of rows} + \text{Total number of collumns} \\ &\qquad + \text{Total number of letters}) \text{ in the Latin square.} \\ &= s + s + s(s-1) \\ &= s(s-1). \end{aligned}$$

$$k = \text{Total number of treatments ina row (or column)or the total number of letters in the Latin square} = s.$$

$$\begin{aligned} r &= \text{Total number of times a treatment appears in all the blocks together} \\ &= \text{Total number of times the corresponding cell appears in all the rows} + \text{in all the columns} \\ &\qquad + \text{Total number of letters contained in that cell} \end{aligned}$$

$= 1 + 1 + (s - 1)$
$= s + 1.$
$\lambda$ = Total number of times a treatment pair appears
$= 1$

Therefore $v = s^2, b = s^2 + s, k = s, r = s + 1$, and $\lambda = 1$. Hence for s =2, 3, 4, 5, 7, 8, 9, 11, 13 etc. we can construct the BIBD's since we can construct the corresponding system of orthogonal Latin squares.

**Example 9.2**

$s = 3.$

There are two orthogonal Latin squares and the superimposed square will be

| | | |
|---|---|---|
| Aa<br>1 | Bb<br>2 | Cc<br>3 |
| Bc<br>4 | Ca<br>5 | Ab<br>6 |
| Cb<br>7 | Ac<br>8 | Ba<br>9 |

The numbers denote the corresponding treatments.

(i) Cell in a row → Treatments in a block. Hence we get the blocks (from rows)

| | | |
|---|---|---|
| 1 | 2 | 3 |
| 4 | 5 | 6 |
| 7 | 8 | 9 |

(ii) Cells in a column → Treatments in a block. Hence (from columns) we get the blocks

| | | |
|---|---|---|
| 1 | 4 | 7 |
| 2 | 5 | 8 |
| 3 | 6 | 9 |

(iii) Cell containing letters → Treatments in a block. Therefore (from Latin letters) we get the blocks

$$\begin{array}{l} A \rightarrow 1\ 6\ 8 \\ B \rightarrow 2\ 4\ 9 \\ C \rightarrow 3\ 5\ 7 \end{array}$$

and (from Greek letters) we get the blocks

$$\begin{array}{l} a \rightarrow 1\ 5\ 9 \\ b \rightarrow 2\ 6\ 7 \\ c \rightarrow 3\ 4\ 8. \end{array}$$

Thus we get 4 sets of 3 blocks each and each set is a complete replication. In general, we get $(s+1)$ sets of s blocks each and each set is a complete replication. Therefore, by Yates series A we construct resolvable BIBD's.

s=4:

$$v = 16, b = 20, k = 4, r = 5, \lambda = 1.$$

$x^2 + x + 1$ is an irreducible polynomial of degree 2 in the mod 2 field and this gives the elements of $GF(2^2)$ as

$$\alpha_0 = 0, \alpha_1 = 1, \alpha_2 = x, \alpha_3 = x + 1.$$

The key line of $L_1$ is $1 + \alpha_{q-1}$ and constructing $L_1$ we can construct the remaining 2 orthogonal Latin squares also. The complete system of orthogonal Latin squares of side 4 is

| | | | |
|---|---|---|---|
| 000<br>(1) | 123<br>(2) | 231<br>(3) | 312<br>(4) |
| 111<br>(5) | 032<br>(6) | 320<br>(7) | 203<br>(8) |
| 222<br>(9) | 301<br>(10) | 013<br>(11) | 130<br>(12) |
| 333<br>(13) | 210<br>(14) | 102<br>(15) | 021<br>(16) |

The corresponding resolvable BIBD will be

| | | | | | |
|---|---|---|---|---|---|
| I: | 1 | 2 | 3 | 4 | from the rows |
| | 5 | 6 | 7 | 8 | |
| | 9 | 10 | 11 | 12 | |
| | 13 | 14 | 15 | 16 | |

| | | | | | |
|---|---|---|---|---|---|
| II: | 1 | 5 | 9 | 13 | from the columns |
| | 2 | 6 | 10 | 14 | |
| | 3 | 7 | 11 | 15 | |
| | 4 | 8 | 12 | 16 | |

| | | | | | |
|---|---|---|---|---|---|
| III: | 1 | 6 | 11 | 16 | from the first set of letters, here numbers |
| | 2 | 5 | 12 | 15 | |
| | 3 | 8 | 9 | 14 | |
| | 4 | 7 | 10 | 13 | |

| | | | | | |
|---|---|---|---|---|---|
| IV: | 1 | 8 | 10 | 15 | from the second set of letters |
| | 4 | 5 | 11 | 14 | |
| | 2 | 7 | 9 | 16 | |
| | 3 | 6 | 12 | 13 | |

| | | | | | |
|---|---|---|---|---|---|
| V: | 1 | 7 | 12 | 14 | from the third set of letters |
| | 3 | 5 | 10 | 16 | |
| | 4 | 6 | 9 | 15 | |
| | 2 | 8 | 11 | 13 | |

This is Yates series A.

Yates series B is obtained from Yates series A by adding one more column with the same treatments in each set of Yates' series A and by introducing a block of the new treatments. Hence in Yates series B,

$$v = s^2 + s + 1, b = s^2 + s + 1, k = s + 1, r = s + 1, \lambda = 1.$$

Hence Yates' series B is a system of symmetric BIBD's.

For example consider $s = 4$. Let the new treatments added be 17, 18, 19, 20, 21. Then we get the following design:

| I: | 1 | 2 | 3 | 4 | 17 |
|---|---|---|---|---|---|
| | 5 | 6 | 7 | 8 | 17 |
| | 9 | 10 | 11 | 12 | 17 |
| | 13 | 14 | 15 | 16 | 17 |

| II: | 1 | 5 | 9 | 13 | 18 |
|---|---|---|---|---|---|
| | 2 | 6 | 10 | 14 | 18 |
| | 3 | 7 | 11 | 15 | 18 |
| | 4 | 8 | 12 | 16 | 18 |

| III: | 1 | 6 | 11 | 16 | 19 |
|---|---|---|---|---|---|
| | 2 | 5 | 12 | 15 | 19 |
| | 3 | 8 | 9 | 14 | 19 |
| | 4 | 7 | 10 | 13 | 19 |

| IV: | 1 | 8 | 10 | 15 | 20 |
|---|---|---|---|---|---|
| | 4 | 5 | 11 | 14 | 20 |
| | 2 | 7 | 9 | 16 | 20 |
| | 3 | 6 | 12 | 13 | 20 |

| V: | 1 | 7 | 12 | 14 | 21 |
|---|---|---|---|---|---|
| | 3 | 5 | 10 | 16 | 21 |
| | 4 | 6 | 9 | 15 | 21 |
| | 2 | 8 | 11 | 13 | 21 |

and the block with the added treatments will be

$$17 \quad 18 \quad 19 \quad 20 \quad 21.$$

Thus we get a symmetric BIBD with

$$v = b = 13, k = r = 4, \lambda = 1.$$

## 9.3 Geometric Method (Finite Projective Geometry)

Any ordered set of N+1 elements $(x_0, x_1, \ldots, x_N)$ where $x_i \in GF(p^n)$ will be called a point of the finite projective geometry of N dimensions, written $PG(N, s), s = p^n$, provided all the $x_i$'s are not zero simultaneously. Two points $(x_0, x_1, \ldots, x_N)$ and $(y_0, y_1, \ldots, y_N)$ are identical if and only if there exists a non-zero element $\alpha$ of $GF(s)$ such that $y_i = \alpha x_i, i = 1,2, \ldots, N$. $GF(s)$ contains $(s-1)$ non-zero elements and so it follows that the same point can be represented in $(s-1)$ different forms corresponding to the values of $\alpha$. Each $x_i$ can take s values. So the ordered set $(x_0, x_1, \ldots, x_N)$ can be chosen in $s^{N+1}$ ways. We have to exclude the case when all $x_i$'s are simultaneously zero if the set is to represent a point. Since each point in this case will be repeated in its $(s-1)$ forms, the number of distinct points in $PG(N, s)$ is $\frac{s^{N+1}-1}{s-1}$.

**Flats**

All the points satisfying a set of $(N-m)$ independent linear equations in $x_i$'s, viz.,

$$\sum_{j=0}^{N} a_{ij} x_j = 0, \quad i = 1,2, \ldots, N-m,$$

where $a_{ij} \in GF(s)$, not all simultaneously zero, is said to form an m-flat or an m-dimensional subspace in $PG(N, s)$. Total number of distinct m-flats in $PG(N, s)$ is given by

$$\phi(N, m, s) = \frac{(s^{N+1}-1)(s^{N+1}-s)(s^{N+1}-s^2)\ldots(s^{N+1}-s^{N-m-1})}{(s^{N-m}-1)(s^{N-m}-s)(s^{N-m}-s^2)\ldots(s^{N-m}-s^{N-m-1})}$$

$$= \frac{(s^{N+1}-1)(s^N-1)(s^{N-1}-1)\ldots(s^{m+2}-1)}{(s^{N-m}-1)(s^{N-m-1}-1)(s^{N-m-2}-1)\ldots(s-1)}$$

$$= \frac{(s^{N+1}-1)(s^N-1)\dots(s^{m+2}-1)}{(s-1)(s^2-1)\dots(s^{N-m}-1)}$$
$$\cdot \frac{(s^{m+1}-1)(s^m-1)\dots(s^{N-m+1}-1)}{(s^{N-m+1}-1)(s^{N-m+2}-1)\dots(s^{m+1}-1)}$$
$$= \phi(N, N-m-1, s).$$

For a 0-flat, N equations are to be satisfied. By convention a 0-flat is taken as a point. The number of 0-flats is equal to

$$\phi(N,0,s) = \phi(N,N-1,s)$$
$$= \frac{s^{N+1}-1}{s-1},$$

i.e. the total number of points in $PG(N, s)$. Since,

$$\phi(N,N,s) = \phi(N,-1,s),$$

an N-flat is defined as a (-1)-flat. The number of (-1)-flats is

$$\phi(N,-1,s) = \phi(N,N,s)$$
$$= \frac{(s^{N+1}-1)\dots(s-1)}{(s^{N+1}-1)\dots(s-1)}$$
$$= 1.$$

The total number of distinct m-flats passing through a fixed set of r points $(r < m)$ is given by

$$\phi(N-r, m-r, s).$$

The total number of distinct points in an m-flat is given by

$$\phi(m,0,s) = \phi(m,m-1,s)$$
$$= \frac{s^{m+1}-1}{s-1}.$$

## FLATS AT INFINITY

The point $(x_0, x_1, \dots, x_N)$ where $x_0 = 0$ is called a point at infinity. It satisfies $N - m = 1$ linear condition, viz. $x_0 = 0$, and hence $m = N - 1$. Therefore it can also be considered as an (N-1)-flat. Hence, it is conventional to choose the (N-1)-flat $x_0 = 0$ as the flat at infinity.

Points in this flat are known as points at infinity and the remaining points are known as finite points. It is easily seen that the number of points at infinity is the same as the number of points in an (N-1)-dimensional geometry, viz.,

$$\phi(N-1,0,s) = \frac{s^N - 1}{s-1}.$$

Hence the number of finite points in $PG(N, s)$ is

$$\phi(N,0,s) - \phi(N-1,0,s) = \frac{s^{N+1}-1}{s-1} - \frac{s^N - 1}{s-1}$$
$$= s^N.$$

Since the $x_0$-coordinate of a finite point is non-zero, we can write it in the unique form $(1, x_1, x_2, \ldots, x_N)$ by dividing throughout by $x_0$.

If all the points of a particular m-flat are points at infinity it will be called an m-flat at infinity. All other m-flats will be called finite m-flats. Since for every m-flat at infinity $x_0 = 0$, there are only $N - m - 1$ other linear equations to be satisfied by the remaining N coordinates which amounts to the definition of an (N-1)-(N-m-1) = m-flat in an (N-1)-dimensional geometry. Hence the number of m-flats at infinity in $PG(N, s)$ is $\phi(N-1, m, s)$ so that the total number of finite m-flats in $PG(N, s)$ is

$$\phi(N,m,s) - \phi(N-1,m,s) = s^{N-m}\phi(N-1,m-1,s).$$

The number of points at infinity in a finite m-flat is

$$\phi(m-1,0,s) = \frac{s^m - 1}{s-1}.$$

Hence the number of finite points in a finite m-flat in $PG(N, s)$ is given by

$$\phi(m,0,s) - \phi(m-1,0,s) = s^m.$$

The number of finite m-flats passing through a given set of $r < m$ finite points is $\phi(N-r, m-r, s)$.

## 9.4 Finite Euclidean Geometry

An ordered set of N coordinates $(x_1, x_2, \dots, x_N)$ where $x_i \in GF(p^n = s)$ will be called a point in N-dimensional Euclidean geometry, written $EG(N, s)$. Clearly, the number of points in $EG(N, s)$ is $s^N$. The set of points satisfying $(N - m)$ independent linear equations of the form

$$a_{i0} + a_{i1}x_1 + \cdots + a_{iN}x_N = 0, \quad i = 1,2, \dots, N - m$$

is said to form an m-flat in $EG(N, s)$. Since these $(N - m)$ equations can be solved in terms of m variables $x_i$, the number of points in an m-flat is $s^m$.

**Relation Between Euclidean and Projective Geometry**

We can establish a one-to-one correspondence between the points in $EG(N, s)$ and the finite points of $PG(N, s)$ as follows:

Since the $x_0$–coordinates of the finite points in $PG(N, s)$ are non-zero, we can write any finite point in the unique form $(1, x_1, x_2, \dots, x_N)$ and this corresponds with the point $(x_1, x_2, \dots, x_N)$ in $EG(N, s)$. By this correspondence an m-flat in $EG(N, s)$ corresponds to a finite m-flat in $PG(N, s)$ and so the number of m-flats in $EG(N, s)$ is $s^{N-m}\phi(N - 1, m - 1, s)$ and the number of points in an m-flat in $EG(N, s)$, is the same as the number of finite points in the corresponding m-flat in $PG(N, s)$, viz., $s^m$. The number of m-flats in $EG(N, s)$ passing through a fixed set of $r < m$ points is the same as the number of finite m-flats passing through those finite points, viz. $\phi(N - r, m - r, s)$.

**Construction of BIBD**

Consider $GF(p^n)$, p prime, and $PG(N, s)$, $s = p^n$ and the following correspondence:

$$\text{Points in } PG(N, s) \rightarrow \text{Treatment in a BIBD.}$$

$$\text{m-flat in } PG(N, s) \rightarrow \text{Block in a BIBD.}$$

Therefore

v = Total number of treatments in a BIBD

= Total number of points in PG(N, s)

$= \phi(N, 0, s)$

$= \frac{s^{N+1}-1}{s-1}.$

b = Total number of blocks in a BIBD

= Total number of distinct m-flats in PG(N, s)

$= \phi(N, m, s)$

$= \frac{(s^{N+1}-1)(s^{N}-1)\ldots(s^{m+2}-1)}{(s^{N-m}-1)(s^{N-m-1}-1)\ldots(s-1)}.$

k = Total number of treatments in a block

= Total numbers of points in an m-flat

$= \phi(N, 0, s)$

$= \frac{s^{m+1}-1}{s-1}.$

r = Total number of times a treatment appears in all blocks

= Total number of times a point appears in all the m-flats

= Total number of m-flats passing through a fixed point

$= \phi(N-1, m-1, s).$

λ = Total number of times a pair of treatments appears in all the blocks

= Total number of m-flats pasing through two fixed points

$= \phi(N-2, m-2, s).$

**Example 9.3**

(i) Let $N = 2, m = 1$.
Then

$$\begin{aligned} v &= s^2 + s + 1, \\ b &= s^2 + s + 1, \\ k &= s + 1, \\ r &= s + 1, \\ \lambda &= 1. \end{aligned}$$

We get Yates series B.

(ii) Let $N = 3, m = 2, \text{and } s = 2$.
We get

$$v = 15, b = 15, k = 7, r = 7, \lambda = 3.$$

We have to consider $PG(3,2)$ and the 2-flats in it. Points in $PG(3,2)$ correspond to treatments and the 2-flats correspond to blocks. Therefore we have to enumerate the 2-flats and the points in them to get the blocks and the treatments in those blocks of the corresponding BIBD.

Any representation of the form $(x_0, x_1, x_2, x_3)$ will be a point in PG(3,2) if all the x's are not simultaneously zero and where the x's are residues mod 2. A 2-flat in PG(3,2) has to satisfy $N - m = 3 - 2 = 1$ equation. Let us enumerate all the 2-flats. The equations are

$$x_i = 0 \quad (4 \text{ equations, therefore } 4 \;\; 2-\text{flats})$$

$$x_i + x_j = 0, \;\; i \neq j.$$

$$\left(\binom{4}{2} = 6 \text{ equations, therefore } 6 \;\; 2-\text{flats}\right)$$

$$x_i + x_j + x_k = 0, \;\; i \neq j \neq k.$$

$$\left(\binom{4}{3} = 4 \text{ equations, therefore } 4 \;\; 2-\text{flats}\right)$$

$$x_0 + x_1 + x_2 + x_3 = 0. \text{ (one equation, therefore one 2-flat).}$$

The first set of 4 2-flats are given by

$$x_0 = 0$$

$$x_1 = 0$$

$$x_2 = 0$$

$$x_3 = 0.$$

The second set of 6 2-flats are given by

$$x_0 + x_1 = 0$$

$$x_0 + x_2 = 0$$

$$x_0 + x_3 = 0$$

$$x_1 + x_2 = 0$$

$$x_1 + x_3 = 0$$

$$x_2 + x_3 = 0.$$

The third set of 4 2-flats are given by

$$x_0 + x_1 + x_2 = 0$$

$$x_0 + x_1 + x_3 = 0$$

$$x_0 + x_2 + x_3 = 0$$

$$x_1 + x_2 + x_3 = 0.$$

The last 2-flat is given by

$$x_0 + x_1 + x_2 + x_3 = 0.$$

The following points satisfy the equation:

$x_0 = 0$:

0001, 0010, 0011, 0100, 0101, 0110, 0111.

$x_1 = 0$:

0001, 0010, 0011, 1000, 1001, 1010, 1011.

$x_2 = 0$:

0001, 0100, 0101, 1000, 1001, 1101, 1101.

$x_3 = 0$:

0010, 0100, 0110, 1000, 1010, 1100, 1110.

The following points satisfy the equation

$x_0 + x_1 = 0$:

0001, 0010, 0011, 1100, 1101, 1110, 1111.

$x_0 + x_2 = 0$:

0001, 0100, 0101, 1010, 1011, 1110, 1111.

$x_0 + x_3 = 0$:

0010, 0110, 0100, 1001, 1011, 1101, 1111.

$x_1 + x_2 = 0$:

0001, 1000, 1001, 0110, 1110, 0111, 1111.

$x_1 + x_3 = 0$:

1000, 1010, 0010, 0101, 1101, 0111, 1111.

$x_2 + x_3 = 0$:

0100, 1000, 1100, 0011, 1011, 0111, 1111.

$x_0 + x_1 + x_2 = 0$:

0001, 1101, 1011, 0111, 0110, 1010, 1100.

$x_0 + x_1 + x_3 = 0$:

0010, 0101, 0111, 1100, 1110, 1001, 1011.

$x_0 + x_2 + x_3 = 0$:

0100, 1010, 1110, 1001, 1101, 0011, 0111.

$x_1 + x_2 + x_3 = 0$:

1000, 0110, 1110, 0101, 1101, 0011, 1011.

$x_0 + x_1 + x_2 + x_3 = 0$:

0011, 0110, 0101, 1010, 1001, 1100, 1111.

Let us denote the 15 points of PG(3,2) by the numerals 1, 2, 3, …, 15 according to the following order:

$$0001 - 1$$
$$0010 - 2$$
$$0011 - 3$$
$$0100 - 4$$
$$0101 - 5$$
$$0110 - 6$$
$$0111 - 7$$
$$1000 - 8$$
$$1001 - 9$$
$$1010 - 10$$
$$1011 - 11$$
$$1100 - 12$$
$$1101 - 13$$
$$1110 - 14$$
$$1111 - 15$$

Then the 15 distinct 2-flats and the points in them (7 points on each) are as arranged below:

| 1 | 2 | 3 | 4 | 5 | 6 | 7 |
|---|---|---|---|---|---|---|
| 1 | 2 | 3 | 8 | 9 | 10 | 11 |
| 1 | 4 | 5 | 8 | 9 | 12 | 13 |
| 2 | 4 | 6 | 8 | 10 | 12 | 14 |
| 1 | 2 | 3 | 12 | 13 | 14 | 15 |
| 1 | 4 | 5 | 10 | 11 | 14 | 15 |
| 2 | 6 | 4 | 9 | 11 | 13 | 15 |
| 1 | 8 | 9 | 6 | 14 | 7 | 15 |
| 8 | 10 | 2 | 5 | 13 | 7 | 15 |
| 4 | 8 | 12 | 3 | 11 | 7 | 15 |
| 1 | 13 | 11 | 7 | 6 | 10 | 12 |
| 2 | 5 | 7 | 12 | 14 | 9 | 11 |
| 4 | 10 | 14 | 9 | 13 | 3 | 7 |
| 8 | 6 | 14 | 5 | 13 | 3 | 11 |
| 3 | 6 | 5 | 10 | 9 | 12 | 15 |

Now, 2-flats correspond to blocks and points correspond to treatments. By this correspondence the above arrangement gives a BIBD with parameters $v = b = 15$, $k = r = 7$, $\lambda = 3$. The numerals denote treatments. This is a symmetric BFBD.

**Construction of a BIBD using Finite Euclidean Gometry**

Here we establish the correspondence as follows:

$$\text{m-flat in EG}(N, s) \rightarrow \text{Blocks of a BIBD.}$$

$$\text{Points in EG}(N, s) \rightarrow \text{Treatments in a BIBD.}$$

Therefore

$$\begin{aligned} v &= \text{Total number of treatments} \\ &= \text{Total number of points in EG}(N, s) \\ &= s^N. \end{aligned}$$

$$\begin{aligned} b &= \text{Total number of blocks in a BIBD} \\ &= \text{Total number of distinct m-flats in EG}(N,s) \end{aligned}$$

$= s^{N-m}\phi(N-1, m-1, s).$

$k =$ Total number of treatments in a block

$=$ Total number of points in a m-flats in $EG(N, s)$

$= s^m.$

$r =$ Total number of times a treatment appears in all the blocks

$=$ Total number of times a point appears in all the m-flats

$=$ Total number of m-flats through a fixed point in $EG(N, s)$

$= \phi(N-1, m-1, s).$

Similarly,

$$\lambda = \phi(N-2, m-2, s).$$

**Examples 9.4**

(i) $N = 2, m = 1.$

$$v = s^2$$

$$b = s\phi(1,0,s) = s\frac{s^2-1}{s-1} = s^2 + s$$

$$k = s$$

$$r = \phi(1,0,s) = s + 1,$$

$$\lambda = \phi(0,-1,s)$$

$$= \phi(0,0,s)$$

$$= \frac{s-1}{s-1}$$

$$= 1.$$

This is Yates series A.

(ii) $N = 3, s = 2, m = 2.$

$$v = s^N = 8$$

$$b = s^{N-m}\phi(N-1, m-1, s) = 14,$$

$$k = s^m = 4$$

$$r = \phi(N-1, m-1, s) = 7,$$

$$\lambda = \phi(N-2, m-2, s) = 3,$$

$$N - m = 3 - 2 = 1.$$

Hence the 2-flats in EG(3,2) have to satisfy one equation. Any representation of the form $(x_1, x_2, x_3)$ will be a point in EG(3,2) where the x's take the values 0 and 1. The independent equations giving the 2-flats are enumerated below:

(1) Equations of the form $x_i = 0,1;\ i = 1,2,3$

| $x_1 = 0,1$ | Gives six 2-flats |
|---|---|
| $x_2 = 0,1$ | |
| $x_3 = 0,1$ | |

(2) $x_i + x_j = 0,1 \quad i \neq j$

| $x_4 + x_2 = 0,1$ | Gives six 2-flats |
|---|---|
| $x_1 + x_3 = 0,1$ | |
| $x_2 + x_3 = 0,1$ | |

(3) The equations $x_4 + x_2 + x_3 = 0,1$ gives two 2-flats.

The points in these 2-flats are as follows:

(1)

<table>
<tr><td>000</td><td>001</td><td>010</td><td>011</td><td rowspan="2">$x_1 = 0,1$</td></tr>
<tr><td>100</td><td>101</td><td>110</td><td>111</td></tr>
<tr><td>000</td><td>001</td><td>101</td><td>100</td><td rowspan="2">$x_2 = 0,1$</td></tr>
<tr><td>010</td><td>011</td><td>110</td><td>111</td></tr>
<tr><td>000</td><td>010</td><td>100</td><td>110</td><td rowspan="2">$x_3 = 0,1$</td></tr>
<tr><td>001</td><td>011</td><td>101</td><td>111</td></tr>
</table>

(2)

<table>
<tr><td>000</td><td>001</td><td>110</td><td>111</td><td rowspan="2">$x_4 + x_2 = 0,1$</td></tr>
<tr><td>010</td><td>011</td><td>101</td><td>100</td></tr>
<tr><td>000</td><td>101</td><td>111</td><td>010</td><td rowspan="2">$x_1 + x_3 = 0,1$</td></tr>
<tr><td>001</td><td>100</td><td>110</td><td>011</td></tr>
<tr><td>000</td><td>100</td><td>011</td><td>111</td><td rowspan="2">$x_2 + x_3 = 0,1$</td></tr>
<tr><td>001</td><td>010</td><td>101</td><td>110</td></tr>
</table>

(3)

<table>
<tr><td>000</td><td>110</td><td>101</td><td>011</td><td rowspan="2">$x_1 + x_2 + x_3 = 0,1$</td></tr>
<tr><td>001</td><td>010</td><td>101</td><td>111</td></tr>
</table>

Let us denote the 8 points in EG(3,2) as follows:

$$000 - 1 \quad 100 - 5$$

$$001 - 2 \quad 101 - 6$$

$$010 - 3 \quad 110 - 7$$

$$011 - 4 \quad 111 - 8.$$

Then the 14 distinct 2-flats and the points on each (4 points) are given below:

| 1 | 2 | 3 | 4 |
|---|---|---|---|
| 5 | 6 | 7 | 8 |
| 1 | 2 | 6 | 5 |
| 3 | 4 | 7 | 8 |
| 1 | 3 | 5 | 7 |
| 2 | 4 | 6 | 8 |
| 1 | 2 | 7 | 8 |
| 3 | 4 | 6 | 5 |
| 1 | 6 | 8 | 3 |
| 2 | 5 | 7 | 4 |
| 1 | 5 | 4 | 8 |
| 2 | 3 | 6 | 7 |
| 1 | 7 | 6 | 4 |
| 2 | 3 | 6 | 8 |

Because of the assumed correspondence the above arrangement gives a BIBD with parameters

$$v = 8, b = 14, k = 4, r = 7, \lambda = 3.$$

# Index